MW01247748

IMISCOE Research Series

This series is the official book series of IMISCOE, the largest network of excellence on migration and diversity in the world. It comprises publications which present empirical and theoretical research on different aspects of international migration. The authors are all specialists, and the publications a rich source of information for researchers and others involved in international migration studies.

The series is published under the editorial supervision of the IMISCOE Editorial Committee which includes leading scholars from all over Europe. The series, which contains more than eighty titles already, is internationally peer reviewed which ensures that the book published in this series continue to present excellent academic standards and scholarly quality. Most of the books are available open access.

For information on how to submit a book proposal, please visit: http://www.imiscoe.org/publications/how-to-submit-a-book-proposal.

More information about this series at http://www.springer.com/series/13502

Gabriel Echeverría

Towards a Systemic Theory of Irregular Migration

Explaining Ecuadorian Irregular Migration in Amsterdam and Madrid

Gabriel Echeverría
International Cooperation Centre
Trento, Italy

University of Trento
Trento, Italy

ISSN 2364-4087 ISSN 2364-4095 (electronic)
IMISCOE Research Series
ISBN 978-3-030-40902-9 ISBN 978-3-030-40903-6 (eBook)
https://doi.org/10.1007/978-3-030-40903-6

This Springer imprint is published by the registered company Springer Nature Switzerland AG.
The registered company address is: Gewerbestrasse 11, 6330 Cham, Switzerland

Para Giacomo

Acknowledgments

The content of this book is the result of a journey which lasted 5 years. As in all journeys, there were times of enchantment and disenchantment, inspiration and frustration, curiosity and tedium, exaltation and discourage. All in all, they were years of privilege, in which I was able to develop, with absolute freedom, an original research project on a subject that is especially meaningful to me.

The research work was made possible thanks to the generous contribution of the FPU Scholarship of the Ministry of Education of Spain and to the support of three institutions, the Complutense University of Madrid, the Instituto Universitario de Investigación Ortega y Gasset and the University of Trento.

The support, help and suggestions of many people have been crucial for the accomplishment of this work. First of all, I would like to thank all the migrants who participated in the research project in Amsterdam and Madrid. Their kindness and generosity were for me a great gift and an example. In particular, I would like to thank Sylvana Cabezas, Ramiro Palacions and their families. They opened me the door to the Ecuadorian migrants' world in Amsterdam and welcomed me as a son. Without their help, this research work would have been impossible. I would also like to thank Lulu Cabezas, Malki, Ñusta and Ñaupac for having hosted me in their house and made me feel at home.

Special thanks to my two doctoral supervisors, Prof. Joaquín Arango and Prof. Giuseppe Sciortino. Their support, orientation and advise were the cornerstone at the base of this project.

I would like to thank Prof. Godfried Engbersen for having hosted me at the Department of Sociology of the Erasmus University of Rotterdam during my research stay in the Netherlands. Thanks also to Willem Schinkel, Arjen Leerkes, Dennis Broeders and all members of the Department for their help and suggestions. A special thanks to Masja Van Meeteren.

The work to transform the doctoral thesis into a book was possible thanks to the support and encouragement of the Centre for International Cooperation in Trento.

Thanks also to the IMISCOE Editorial Committee for selecting my proposal for the IMISCOE Springer Competitive Call for Book Proposals and to Anna Triandafyllidou for her support. Many thanks to Evelien Bakker, Bernadette

Deelen-Mans and Alexandre James at Springer and, of course, to the two anonymous reviewers, whose constructive and detailed feedback helped to refine the overall argument of the book.

I had the privilege to discuss parts of this work with a number of scholars and researchers. Thanks to Claudia Finotelli, Paolo Boccagni, Sandro Mezzadra, Sébastien Chauvin, Nicholas De Genova, MariaCaterina La Barbera, Sarah Spencer, Refugio Chávez and Rosa Aparicio.

A great part of the sorrows and joys of this journey was shared with my doctorate colleagues. Thanks to Adriana, Aurelis, Carolina, Elisabeth, Inara, Damián and Joaquín. A special thanks to Elena with whom I spent endless hours at the library and on the phone.

This book would not have been possible without the generous and expert help of Irene Diamond who meticulously reviewed the English of each page.

During these years, three people who had a crucial influence on my life and on the ideas here discussed passed away. I would like to remember my grandfather, Bolívar Echeverría Paredes; my uncle, Bolívar Echeverría Andrade; and Gabriele Viliani.

It is not necessary, unless for the pleasure to do it, to mention the gratitude to my dear families in Ecuador and Italy. A big thanks to my parents. If it is true that in order to have a pleasant life it is important to choose them well, I recognize myself this merit. All my love to my sisters and brothers, Maria Chiara, Francisca and Tomás. Infinite gratitude to Adriano and Natalia.

This book is dedicated to Giacomo and the joyful curiosity in his eyes. Nothing of all this would have been possible without you, Dafne.

Contents

Chapter 1
Introduction

Palomar s'è distratto, non strappa più le erbacce,
non pensa più al prato: pensa all'universo.
Sta provando ad applicare all'universo
tutto quello che ha pensato del prato.
L'universo come cosmo regolare e ordinato
o come proliferazione caotica.
L'universo forse finito ma innumerabile,
instabile nei suoi confini,
che apre entro di sé altri universi.
L'universo, insieme di corpi celesti, nebulose,
pulviscolo, campi di forze, intersezioni di campi,
insiemi di insiemi... (Mr. Palomar's mind has wandered, he has stopped pulling up weeds. He no longer thinks of the lawn: he thinks of the universe. He is trying to apply to the universe everything he has though about the lawn. The universe as regular and ordered cosmos or as chaotic proliferation. The universe perhaps finite but countless, unstable within its borders, which discloses other universes within itself. The universe, collection of celestial bodies, nebulas, fine dust, force fields, intersections of fields, collections of collections...)

Italo Calvino

Irregularity is a juridical status that describes the relation between a migrant and one or more states. As a social phenomenon, it does not derive from the migrations themselves, rather, it is the result "of the existence of a structural tension between the social preconditions and the political preconditions" that support them (Sciortino, 2007). The social space, following this interpretation, is the scenario where two different and opposed logics interact. On the one hand, there is the logic of free movement of people and goods that is favoured by socio-economic forces like the market-economy, globalization or transnationalism. On the other hand, there is the logic of the states, political-juridical constructions, historically and ideologically differentiated, that claim the power to delimit the space and to regulate the movement of factors across it. Irregularity would then be the result of the clash between these two logics that determine a numerical difference between the migrants that

G. Echeverría, *Towards a Systemic Theory of Irregular Migration*, IMISCOE Research Series, https://doi.org/10.1007/978-3-030-40903-6_1

move across the geographical space, established by the first logic, and the migrants who are allowed to do that, established by the second logic.

The divergence between these two logics has become particularly relevant in the present age of globalization. In the previous historical phase, the "social space", understood as the space within which the majority of social transactions take place, tended to better overlap with the "political space", understood as the space where those transactions are regulated by a sovereign power. In that context, the main social interactions occurred within the boundaries of the states and those that crossed frontiers were rather limited and then more easily controllable. Human mobility, which is, of course, not a novelty of the current historical moment, took place massively before globalization, but it largely occurred through channels established by the states and often under their own auspice.

The growing liberalization in the exchange of goods, capital and information, as well as the drastic reduction in the costs of, and time needed for, the exchanges, in other words globalization, have determined a dramatic change in the previous patterns of mobility. Indeed, the fast and worldwide development of interconnections between individuals and societies has led to an inversion in that overlapping tendency between the "social space" and the "political space". This process has uncovered, once more, the possible conflict between the inner logic of the each space. In a certain sense, it could be said that globalization is determining a spill-over of the "social space" beyond the boundaries of the "political space" as was prefigured by the modern national state. Faced by these phenomena, states have reacted in a differentiated way. On the one hand, they seem to be ceding sovereignty as regards the circulation of goods, capital and information. On the other, however, they seem to be widely opposed to the free circulation of people. The contradiction between these two tendencies has been successfully summarized by James Hollifield's image of a clash between markets and states (Hollifield, 2000). Focusing on the effects of this conflict over migrations, Douglas Massey highlighted the existence of a "postmodern paradox" because it is possible to see at work at the same time "global forces" and "restrictive policies" (Massey, 1999).

It is within the frame of this paradox that irregularity can be better understood. Social forces seem to be pushing for a greater mobility of peoples across the globe, while political forces try to regulate or stop such movement. The mismatch between the fluxes generated by the former and those accorded and legitimated by the latter determines that a consistent number of migrants move, reside and work irregularly.

If, in an abstract manner, irregularity can be explained as the result of this conflict, reality, as always happens, provides a more complex scenario where a number of different factors have to be considered and where the role of the actors (e.g. states, migrants, capital, etc.) is more ambiguous and less decisive than it may appear at first sight.

The growing impact of irregular migration in receiving countries in the last few decades, in spite of the efforts against it taken by the states, has fostered anxieties in the public opinion of the latter and attracted the attention of the scientific community (Arango, 2013; Broeders, 2009). From the mid-seventies in the United States and the early nineties in Europe, the study of irregularity and control policies by the states has produced a great variety of interpretations and analyses. These, from a diversity of perspectives, have tried to answer four fundamental questions: A. How can irregular migration be explained? B. What determines the failure or low efficacy of the control policies? C. What are the main impacts of irregular migration on the receiving societies? D. How do irregular migrants manage to live in a supposedly hostile environment? What strategies do they develop? What abuses do they suffer?

Answers, as highlighted by many authors, have been in general partial, if not inadequate, in their explicative capacity. This was often the consequence of over-simplistic analyses or mono-causal argumentations. The criticism has discovered that the reasons for this problems lie both in the lack of theoretical ambition and in the scarcity of empirical evidence (Baldwin-Edwards, 2008; Bommes, 2012; Cvajner & Sciortino, 2010; Düvell, 2006). This two different perspectives, the theoretical and the empirical, moreover, have often operated without establishing an effective dialogue with one another.

This study has started precisely from the two elements that have emerged from this brief discussion. On the one hand, irregular migration represents an extremely interesting phenomenon, one that particularly reveals the dynamics, conflicts and contradictions of our age. As pointed out by McNevin: "perhaps more than any other cross-border flow, irregular migration captures the symbolism of borders under siege in an age of globalization" (McNevin, 2009, p. 168). On the other hand, the comprehension of irregular migration still presents a number of limitations. Yet, the research on irregular migration does not aim at being simply a way to elucidate the particular aspects of a specific social phenomenon, but rather to provide a viewpoint from which to observe the structure and dynamics of contemporary society as a whole. In this sense, through the study of irregular migration, this book aspirates to contribute, with the highest humility, to the greatest task for every generation of researchers, the comprehension of "the spirit of their time".

1.1 Research Questions and Design

There have been two driving forces that have sustained and fostered the research work at the base of this study: firstly, the curiosity for a phenomenon, irregular migration, that is emblematic of the contradictions and complexities of the age of globalization and then the dissatisfaction with most of the available explanations.

The curiosity was not so much aroused by the scenes of the overcrowded boats trying to cross the Mediterranean or of the people jumping over the fences in Tijuana, in order to achieve their "American dream". After all, a great deal of human history has been about people trying to overcome barriers, no matter whether they are geographical or political, in order to improve their living conditions. What really intrigued me was on the other side of those barriers. Why were the rich states that cried against the "invasion", with all their armies, resources and technologies, still unable to stop these hordes of "miserable" people? Was it possible that after four centuries of adjustments and rethinking, the state, epitome of modern politics, had not yet been able to solve the most elemental problem, that of populations coming and going? How could irregular migrants live, work and fulfil their dreams within societies that, at least in principle, refused their presence? Irregular migration appeared to me as a captivating phenomenon because it evidenced the incongruence between the idea of states as the all-embracing, all-mighty controllers of socio-political interactions, and a much more complex and thriving reality made up of conflicts, ambivalences and uncertainty. Reflecting and researching on irregular migration, from this point of view, seems to me not simply a way to elucidate the particular aspects of a specific social phenomenon, but rather to provide a viewpoint from which to observe the structure and dynamics of contemporary society as a whole in the current age of globalization.

A preliminary review of the literature on irregular migration provided me with a large number of different, often contrasting, answers. Depending on the point of view, scholars and researchers had explained the phenomenon as the results of disparate causes, such as: the weaknesses of states, the ability of migrants, the interests of capitalists, the support of criminal networks, etc. As I proceeded in the exploration, I found myself in the paradoxical situation of becoming more and more fascinated by the new approach I found, and, at the same time, more frustrated by the incongruence of the complex puzzle that was emerging. Furthermore, it appeared that each theorization effort usually emerged from the analysis of a particular national case. Thus, for instance, if in a certain place, the role of efficient smuggler networks had been crucial, irregular migration had to be explained everywhere as the result of smuggler networks. Besides, since the studied cases were rather limited, these mono-causal, undifferentiated explanations were generalized without a solid empirical control base. What seemed to be missing, then, was broader and more systematic work of comparison, in other words, one that made it possible to assess similarities and differences between different cases and therefore to offer material for the development of a more general and sophisticated understanding of irregular migration.

On the basis of these initial reflections, I decided to start this work with a very broad and general research question in mind: How can irregular migration be explained? A twofold strategy was formulated in order to add a grain of sand to the building of a better understanding of this phenomenon.

On the one hand, a theoretical study was developed. The objective of this study was to critically analyse the different theories that have been proposed to explain irregular migration and to prepare an alternative theoretical framework. The main research questions of this study were: what it is known about irregular migration? What have been the main theoretical explanations of the phenomenon? What are the strengths and weaknesses of such explanations? Is it possible to find an alternative theoretical framework that is able to reconcile the strengths and overcome the weaknesses of the other theories? Building on the critiques to the principal theoretical explanations of irregular migration, the study focused on the theoretical work of Niklas Luhmann in order to search for a more effective theoretical framework. This approach helped to overcome a number of theoretical difficulties that have characterized this field of research. For instance, it was possible to go beyond a dichotomist understanding of the relation between agency/structure and to retrieve a social perspective where a statist one had been clearly dominant. The result was the elaboration of an analytical framework that enabled the possibility of linking the social characteristics of the irregular migration phenomenon to the structural features of the considered contexts, as well as the understanding of irregular migration as a systemic and differentiated phenomenon.

On the other hand, an empirical study was developed. The objective of this study was to compare the experience of irregular migrants in two receiving contexts and to assess the differences and similarities that characterized the two cases. The aim was to offer empirical material for the theoretical reflection. The chosen case was that of Ecuadorian irregular migrants in the cities of Amsterdam and Madrid. This choice responded to two main explanations. The Ecuadorian migration phenomenon, because of its relatively time circumscribed characteristics and its economical motivations, appeared particularly appropriate for a "at destination" comparative research. Migrants in the two receiving contexts could be considered reasonably similar. In addition, the two cities, while having enough elements in common to avoid the risk of comparing "oranges and apples", were at the same time very different. This allowed for a "most different cases" research strategy, which appeared particular stimulating for theory testing and possible extension.

The empirical study consists of two parts. First, a context study was developed, which comparatively analysed the main structural characteristics of the two cities. Then, a fieldwork that combined ethnography and the collection of 30 in-depth interviews with irregular migrants in each context was developed. The main research questions that prompted this study were: What have been the main structural characteristics affecting migration in the two contexts (migration history, migration regime, economics, welfare state typology, public and political opinion)? What has been the experience of Ecuadorian irregular migrants within the two different contexts? What have been the most important differences and similarities? In particular: what have been the main legal trajectories developed by the migrants within the two contexts? What has been their experience regarding the work sphere (sectors, conditions, controls)? What has been their experience of state controls? Finally, what was their experience regarding basic life facets such as housing and healthcare access?

Although the theoretical and the empirical studies can be considered as separate entities and each has a certain degree of autonomy, they were actually developed together and imagined as complementary parts of a single research effort. Following Derek Layder's "adaptive theory" methodology (Layder, 1998), a purely inductive or a purely deductive approach was avoided. Instead, an attempt was made to establish a permanent dialogue between the theoretical and empirical parts of this study. Adaptive theory focuses on the construction of novel theory by utilizing elements of prior theory (general and substantive) in conjunction with theory that emerges from data collection and analysis. It is the interchange and dialogue between prior theory (models, concepts, conceptual clustering) and emergent theory that forms the dynamic of adaptive theory (Layder, 1998, p. 27). The results that gradually emerged from the empirical work in this study influenced the theoretical reflections while, at the same time, the concepts and ideas emerging from the theoretical work helped to orient and improve the empirical work.

In the concluding part, then, the initial and more general research questions – what is irregular migration and how is it possible to explain it? – were raised again. Combining the results of the contextual study and the fieldwork concerning Amsterdam and Madrid, an attempt was made to establish possible relations between the structural characteristics of the two contexts and the different irregular migration realties that emerged within them. As a result, a differential, systemic explanation of irregular migration was proposed, and its advantages, in comparison with more "orthodox" explanations, were discussed. Finally, combining the results of the theoretical and empirical studies, and by means of a process of abstraction, a preliminary theoretical typology of irregular migration realities in relation to the structural characteristics of the contexts was suggested.

1.2 Structure of the Book

The book is divided into three main parts. In the first part, the results of the bibliographical and theoretical study are presented. In Chap. 2 the existing literature on irregular migration is examined, identifying the main topics, lines of inquiry and scientific debates. Chapter 3 contains a critical analysis of the different theoretical approaches that have been developed towards an understanding of irregular migration. In Chap. 4 an alternative framework for the theoretical understanding of irregular migration, based on the works of Niklas Luhmann's social systems theory, is proposed.

In the second part, the results of the empirical study are portrayed. In Chap. 5 the empirical research design and methodology are discussed. In Chap. 6, a comparative analysis of the main structural characteristics of the cities of Amsterdam and Madrid is made. Chapter 7 deals with an elaboration and discussion of the results of the fieldwork on the experience of Ecuadorian irregular migrants in Amsterdam and Madrid.

In the third and concluding part of the book, a "dialogue" between the previous two parts is attempted and some further steps towards a systemic theory of irregular migration are proposed. Chapter 8 provides a discussion regarding the results of the empirical study, comparing the explicative capacity of the "classical" theoretical approaches, analysed in Chap. 3, and of the systemic approach, developed in Chap. 4. On the basis of this discussion and its results, in the concluding part, a systemic analytical framework of irregular migration is proposed and the strengths and weaknesses of the book are discussed.

Bibliography

Arango, J. (2013). *Exceptional in Europe? Spain's experience with immigration and integration.* Washington DC: Migration Policy Institute.
Baldwin-Edwards, M. (2008). Towards a theory of illegal migration: Historical and structural components. *Third World Quarterly, 29*(7), 1449–1459.
Bommes, M. (2012). Illegal migration in modern society: Consequences and problems of national European migration policies. In C. Boswell & G. D'Amato (Eds.), *Immigration and social systems. Collected essays of Michael Bommes* (pp. 157–176). Amsterdam: Amsterdam University Press.
Broeders, D. (2009). *Breaking down anonymity digital surveillance on irregular migrants in Germany and the Netherlands.* Rotterdam, The Netherlands: Erasmus Universiteit.
Cvajner, M., & Sciortino, G. (2010). Theorizing irregular migration: The control of spatial mobility in differentiated societies. *European Journal of Social Theory, 13*(3), 389–404.
Düvell, F. (2006). *Illegal immigration in Europe: Beyond control?* Basingstoke, UK/New York: Palgrave Macmillan.
Hollifield, J. (2000). The politics of international migration. In C. Brettell & J. Hollifield (Eds.), *Migration theory: Talking across disciplines* (pp. 137–185). New York: Routledge.
Layder, D. (1998). *Sociological practice: Linking theory and social research.* London: Sage.
Massey, D. S. (1999). International migration at the dawn of the twenty-first century: The role of the state. *Population and Development Review, 25*(2), 303–322.
McNevin, A. (2009). Contesting citizenship: Irregular migrants and strategic possibilities for political belonging. *New Political Science, 31*(2), 163–181. https://doi.org/10.1080/07393140902872278
Sciortino, G. (2007). Algunos elementos para comprender a los "irregulares". *Vanguardia Dossier, 22*, 106–109.

Part I
Theoretical Study

Chapter 2
The Study of Irregular Migration

The study of irregular migration as a specific social phenomenon took off during the 70s in the US. Since then, the academic interest has continually grown and spread, first to Europe and, in the last years, to other regions worldwide. This interest can certainly be related to the increasing attention paid to the study of migrations more in general (Castles & Miller, 1993). The trend can be linked to those broad and complex social and economic changes, often subsumed under the concept of globalization. The specific focus on irregular migration, though gaining momentum throughout the 1980s, reached preeminent attention in the 1990s. On both sides of the Atlantic, the explosion of the so-called "migration crisis" (Zolberg & Benda, 2001) and the emergence of irregular migration as a widespread social fact raised the attention of public opinion and academics alike. Moreover, in recent years, what seemed at first to be an issue concerning only the high-income regions of the planet, now involves also medium and low-income ones, making irregular migration a truly global structural phenomenon (Cvajner & Sciortino, 2010a; Düvell, 2006).

Accordingly, after a lapse of two decades, a topic that for a long time had been relatively marginal (Anderson & Ruhs, 2010; Bloch & Chimienti, 2011) became the object of numerous studies and of a consistent and diversified literature. Given the complexity of the phenomenon, its multiple dimensions, and levels of social interaction, its study has inevitably taken a multidisciplinary path. The literature has rapidly expanded in many directions and, today, irregular migration constitutes an important subfield within migration studies.

This chapter will present a general overview of the main directions and developments that the research on irregular migration has taken. Even if some scholars have lamented a limited cumulative effort, many studies are available and it is now possible to refer to them as a solid starting point for analysis. To avoid getting trapped in difficult and sometimes redundant disciplinary distinctions, this overview will focus on the key issues that have been researched from different perspectives. In this respect, it seems possible to identify six main general thematic fields.

G. Echeverría, *Towards a Systemic Theory of Irregular Migration*, IMISCOE Research Series, https://doi.org/10.1007/978-3-030-40903-6_2

2.1 Definition and Taxonomies

Since migration is a complex and multifaceted social phenomenon, an important and on-going debate has focused on terminology issues. Many terms and definitions have been proposed: irregular, illegal, undocumented, clandestine, unauthorized, informal, unregistered, *sans papier*, etc. (Baldwin-Edwards & Kraler, 2009; Düvell, 2006; Jordan & Düvell, 2002; Triandafyllidou, 2009; Vasta, 2011). Each of these has a different focus or emphasis, as well as some advantages and problems.

As pointed out by Nicholas De Genova, the choice of a term does not occur in a social vacuum and it is not politically neutral, for this reason, it should therefore not be taken uncritically (De Genova, 2002). In fact, within a field that has become increasingly politicized, it is not surprising that the selection of terms has assumed a contested nature (Anderson & Ruhs, 2010). Researchers have had to deal not only with classic epistemological problems of definition and perspective but also with the social meaning and connotations that the different terms have in specific contexts. Especially in the last decade, this issue has become increasingly problematic. A number of negative social myths and stereotypical images, usually associated with crime (Castles, 2010; Coutin, 2005b; Dal Lago, 2004; Koser, 2010), have been connected to irregular migration in the public debate and media (Van Der Leun, 2003; Van Meeteren, 2010). Perhaps the most heated dispute has surrounded the use of the term "illegal migration". On the one hand, some scholars have considered that the use of this term contributes to the negative social myths (Koser, 2010; Schrover, Van Der Leun, Lucassen, & Quispel, 2008) and has a criminalizing effect (Düvell, 2006). In a similar fashion, others have stated that its use is simply incorrect since an act can be illegal, whereas a person cannot be so (Castles, 2010; Schinkel 2005 in Engbersen & Broeders, 2009). On the other hand, some scholars have alleged that the term must be used, but in a critical way. From this perspective, it is precisely the process of social and political construction of "illegality" and its consequences that needs to be researched, in particular, the way in which migrants become "illegal migrants" (De Genova, 2004; Willen, 2007).

Behind this terminological debate, there is hidden a related, more substantial one, which is conceptual. Whatever term is adopted, two questions need to be addressed: (a) to what phenomenon does it refer?; (b) from whose perspective?

A certain but far from unanimous consensus has been reached about the fact that the term should refer to the relations between a migrant and a set of rules established by the state, and not to a migrant him/herself (De Genova, 2002). Irregular migration would then be the outcome of the interaction between human mobility across social spaces and the enactment of policies within those very same spaces. In this sense, "the adjective, irregular, does not belong to the domain of description of the migration flows, but only to their interactions with political regulations" (Sciortino, 2004b, p. 21). The complexity of the first question becomes evident once it is recognized that there are many possible types, degrees and dimensions of irregularity (Baldwin-Edwards & Kraler, 2009; Broeders & Engbersen, 2007; Düvell, 2011c; İçduygu, 2007; Triandafyllidou, 2009; Williams & Windebank, 1998).

The term, in fact, can refer to migrants' non-compliance with the rules of entry, residence, employment or a combination of these (Van Der Leun, 2003; Van Meeteren, 2010); to a number of legal statuses implying very different social and economic conditions (Chavez, 1991; Massey et al., 1998; Van Nieuwenhuyze, 2009); and to different forms of social stratification and hierarchy (Castles, 2004; Cvajner & Sciortino, 2010b; López Sala, 2005; Sciortino, 2013; Vasta, 2011). Status, moreover, is not as clear-cut as one might expect, and there is room for forms of legal ambivalence, semi-legality, legal illegality, and formal informality (Düvell, 2011b). Furthermore, "behind the notion of irregular migration there is today a set of interpretative frames, stereotypes, folk wisdom, icons and slogans that makes it a part of a complex symbolic discourse" (Cvajner & Sciortino, 2010b, p. 390). Finally, "the meaning of irregularity shifts across time and space, it is a fluid construction" (Schrover et al., 2008, p. 10); "It is not an "on-off" condition, but rather a bundle of statuses variously significant in different contexts" (Ruhs & Anderson 2006 in Bommes & Sciortino, 2011, p. 219). Depending on where migrants are, they can move in and out of irregularity (Reyneri, 1998), in different ways (Van Der Leun, 2003), for longer or shorter periods. States, on their side, can turn irregular migrants into legal foreign residents, or the other way around "with the single stroke of a pen" (Sciortino, 2013).

In an attempt to organize this diversity of possibilities, once the aspiration for a clear-cut, yes or no, all-embracing definition was abandoned, two main paths have been followed. The first has aimed to develop more flexible definitions, in order to see irregularity not as a fixed status but as a process (Bloch & Chimienti, 2011; Castles, 2010; Jordan & Düvell, 2002). From this perspective, it has been proposed to go beyond the illegal/legal divide and, instead, to understand irregularity as a particular set of conditions within a continuum between two ideal types. On the one hand, there is the "total irregular" (entry, residence, work, illegal practices) and, on the other, "the perfect citizen", somewhere in between all the different cases of "semi-compliant" migrants (Bridget Anderson & Ruhs, 2010). In a similar fashion, irregularity has been defined as an "in-between state among regularisability and deportability" (Garcés-Mascareñas, 2012).

The second path has been to develop taxonomies of different types of irregularity (Bloch & Chimienti, 2011; Haidinger, 2007). Many criteria have been used: ways into irregularity (irregular border-crossing, visa overstaying, refused asylum-demand, violation of the obligation to leave the territory, ineffective deportation, bureaucratic failure/befallen irregularity, birth from irregular parents); duration of stay (limited-, circular- or settlement-irregular migration); types of law violation (irregular entry, residence or work); channels and motivations (smuggled, trafficked, voluntary or forced irregular migrations); irregular migration composition (family, refugees or labour irregular migrations) (Düvell, 2011b; Koser, 2010; Sciortino, 2004b; Vogel & Cyrus, 2008). Regarding the different ways in and out of irregularity and, in order to capture the diversity of possibilities, what has recently been proposed is the distinction among geographical or migration flows, demographic flows and status-related flows (Kraler & Reichel, 2011).

The choice of a certain term implies also the adoption of a specific point of view and of a certain "subjective" perspective. Though this is inevitable, it is important to bear it in mind at every stage. From this perspective, it is possible to distinguish between both taxonomies from above (i.e. the state's point of view) and from below (i.e. the migrant's point of view) (Bloch & Chimienti, 2011). Many scholars have discussed how the term illegal migration entails the adoption of the point of view of the state, which tends to interpret the phenomenon as problematic and challenging (Frank 2008 in Anderson & Ruhs, 2010). This point echoes a more general epistemological and methodological critique of what has been defined as "methodological nationalism" (Castles, 2010; Mezzadra, 2011; Wimmer & Glick-Schiller, 2003). The uncritical adoption of a terminology developed within a statist paradigm, it is argued, leads to distorted representations and to the misperception of a "constructed reality" as if it were the natural one. In this regard, it is important not to forget that much of the terminology used to address issues relating to migration has been developed by state administrations in order to deal with these very issues. Van der Leun, recalling the work of Scott, has warned against those "state simplifications" that are produced and continuously refined to classify migrants (Scott, 1998; Van Der Leun, 2003).

An interesting distinction that the term illegal is unable to capture is the one between what is considered legitimate by the state ("legal") and what is legitimate for people ("licit"). Many trans-border movements of people are illegal because they defy authority, but they are quite acceptable, "licit", in the eyes of participants. Since the state controls those who occupy, use or cross its territory, individuals who contest or bypass controls are bringing into discussion the legitimacy of the state, by questioning its ability to control its territory (Schrover et al., 2008). This example shows the possible conflict between the legal and the political terminologies. Furthermore, if it is considered that, in every national context different legal and political cultures, ideas of national identity, and perceptions of migration are at work (Boswell & D'Amato, 2012; Düvell, 2011b; Kraler & Rogoz, 2011), a full picture of the complexity surrounding the definition and social meaning of irregular migration becomes evident.

The term that has been chosen for this study is irregular migration. Even if this term is not free from possible critiques, its extensive use, especially in the European literature, its flexibility and its relatively neutral perspective make it a suitable tool.

2.2 Irregular Migration from an Historical Perspective

Even if the interest surrounding irregular migration was only aroused in the 1970s, the phenomenon certainly did not appear then. An important line of research has investigated the historic origins and evolution of irregular migration. This task has produced two main types of research. On the one hand, there have been more general accounts on the origin, evolution and trends of irregular migration (Garcés-Mascareñas, 2012; Hollifield, 2004; Schrover et al., 2008; Sciortino, 2004b; Torpey,

1998). On the other, there have been more specific, case-centred studies that enquired into the reasons, ways and moments in which irregular migration appeared in different regions or countries throughout the world. These efforts led to the development of specific national studies and, to a lesser extent, in recent years, to a number of international comparative studies.

If irregular migration is the result of the interaction between migrations and state enforcement of controls over migrants, the history of irregular migration "coincides with the history of attempts by states to gain control over the composition of their population" (Sciortino, 2013). It was the attempt by states to "monopolize the legitimate means of movement" (Torpey, 1998) that made irregular migration emerge as a correlated by-product. Yet, if it is true that, as Sciortino citing Paul of Tarsus has pointed out, "where there is no law, there is no violation" (Sciortino, 2013) it is also true that the existence of a law does not automatically imply its violation. In this sense, the history of irregular migration is not simply the story of migration controls and their implementation, but the story of the interplay of the latter with actual migrants. From this perspective, although the conflict between controls and migrations occurred in a differentiated manner throughout history and geography, and even today there is not one single picture, four broad historical phases seem to be discernible.

The first phase goes from the moment in which nation-states started to coalesce as the main form of political organization, in the sixteenth century, to World War I. This period was mostly marked by weak controls and unrestricted migrations. The old forms of political, territorial and population control were slowly transformed into new, statist ones. National borders became more important than other territorial boundaries, such as the municipal ones (Fahremeir 2007 in Schrover et al., 2008). The process was driven by the diffusion of nationalist ideologies and the idea that a specific population corresponds to each state. Along these lines, states started to develop both legal and administrative mechanisms to register and control their populations, to regulate their borders, and to manage foreign populations (Torpey, 1998). Although instruments to control the movement of vagrants, poor foreigners or unwanted populations (for ethnic, racial or religious reasons) had previously existed at a local level in many contexts (Schrover et al., 2008; Sciortino, 2013, 2017; Zolberg, 2003), "the idea that spatial movements should be considered primarily in terms of their having complied, or failed to comply, with a certain set of generalized, abstract regulations" emerged only at this time (Sciortino, 2013).

Nevertheless, the effective ability to enforce this idea was slowly and unevenly accomplished (Torpey, 1998). For a long time, the ability of state to "effectively control the legitimate means of movement" was still in an embryonic phase. At the same time, although its characterization as a laissez-faire era is certainly overstated, this period can still be considered one of relative openness. The point is not that there were no controls or restrictions at all, but that, since there was a certain equilibrium between the need of migrants in certain societies and overpopulation in others (Hollifield, 2004; Torpey, 2000), migrations were habitually welcomed. The combination of these two circumstances, the embryonic condition of immigration controls and the welcoming character of immigration fluxes made irregular migration quite a marginal if not negligible phenomenon (Hollifield, 2004).

The second phase corresponds approximately to the interwar period. This phase can be regarded as one of increased controls and limited unwanted migrations. States came close to realizing "the bureaucratic fantasy of achieving total control over society" (Ronsenberg, 2006, p. 7 in Schrover et al., 2008). Both their ability and aspiration to control populations were prompted by a number of factors. On the one hand, the material possibilities of states increased thanks to the technological and economic improvements brought by scientific and industrial revolutions. This led to the creation of large and effective bureaucracies capable of regulating and conditioning most social transactions (Garcés-Mascareñas, 2012). The identification and registration of populations were seen as the first steps in order to be able to "read" societies, "embrace" them and make surveillance effective (Broeders, 2009; Scott, 1998; Torpey, 1998). By the 1920s, "the legal and administrative apparatus able to distinguish between citizens and foreigners and, within the latter category, between lawful and unlawful residents" (Sciortino, 2013, p. 6) had become widespread. In this sense, "the urge to control became the ability to control" (Schrover et al., 2008, p. 16).

On the other hand, this period was characterized by the strong affirmation of nationalism, often conflated with racist and xenophobic ideas (Brubaker, 1992; Hobsbawm, 2012). The main consequence was a restrictive turn against migrations that was firstly enacted by the US (Ngai, 2014) and then by most of the other receiving countries (Baldwin-Edwards, 2008; Lucassen & Lucassen, 2005; Schrover et al., 2008). This second period saw the concomitant rise of controls and a decrease in international migration. Whereas the two factors are certainly related, the reduction of international fluxes also had other explanations, mostly related to the changed conditions in the sending countries. In this context of increased control-competency and diminished migratory pressures, irregular migration remained a minor phenomenon.

The third period goes from the end of World War II to the 1970s. This phase can be characterized by a further increase in the control capacity of states, accompanied, however, by a high demand for foreigners. In the aftermath of the war, the demand for workers rapidly increased in Northern European countries and in the US. As pointed out by Baldwin-Edwards, the types of migration varied according to historical, cultural and geographical parameters. The classic immigration settler societies chose permanent immigration over labour migration; postcolonial countries opted for inflows of their colonial citizens; other countries, like Germany, Austria, Switzerland and Belgium, relied on temporary labour schemes (Baldwin-Edwards, 2008). While this categorization describes preponderant patterns, most countries combined strategies and shared the illusion of "migration management" (Garcés-Mascareñas, 2012; Ngai, 2014). Since the priority was to fulfil the demands of a booming economy, those migrants that were able to enter the countries irregularly and found employment were usually and tacitly regularized. Thus, irregular migration was not considered a major problem but rather a transitional phase in the path of migrants. "Expulsion, albeit formally a generalized sanction for irregularity, was mainly interpreted factually as a selective measure to deal with foreign misfits and troublemakers" (Sciortino, 2013, p. 6).

The fourth period goes from the 1970s to our days. Increasingly conflictive relations between receiving states and migratory pressures have characterized this phase. The combination of powerful control systems and masses of migrants willing to travel and, finally, able to do so, set the conditions for irregular migration to become a widespread and sizable phenomenon in unprecedented terms. This explains the vast attention that this period has received and the production of an extensive literature. Scholars have widely discussed the complex structural changes that have occurred in industrialized countries since the mid-1970s. These changes, often referred to as "the end of Fordism", "the rise of the post-industrial economy", or "the economic restructuring" have greatly affected the productive organization, the labour-market structure, and labour relations in the receiving societies (McNevin, 2009; Mezzadra, 2011; Morokvasic, 1993; Piore, 1980; Sassen, 1996; Schierup, Hansen, & Castles, 2006; Wallerstein, 2004).

The impact of these changes has had a long-term effect on the approach to migration and on its management. The turning point was the 1973 oil crisis which implied the abrupt end of the recruitment programmes and the setback of the tolerant and flexible attitude towards irregularity (Engbersen & Van Der Leun, 2001; Zolberg, 2003). It was at this point that the unintended effects of migration policy became manifest, with migration increasingly dealt with as a problem rather than as a resource (Arango, 2005; Broeders, 2009; Sciortino, 2000; Van Meeteren, 2010). The idea that migrants could be used as a commodity in the productive process proved false. Migrants had no intention to return to their countries of origin. Moreover, they had acquired a full set of rights that entitled them to benefit from welfare state provisions, to reunify their families, and to eventually become citizens. Besides, they had developed their own businesses and networks. All this implied that, once started, migrations displayed a self-sustaining dynamic, largely independent from political decisions (Massey, 1999). "The response to this perceived threat has been a building up of visible and invisible walls" (Garcés-Mascareñas, 2012, p. 23). The goal was not only to prevent new entries but also to shelter the welfare state and make access to rights increasingly complicated. As mentioned, the idea that states had lost control over migration became popular both among politicians and the public opinion and started to produce long-lasting effects. Consequently, despite the economic recovery and the renewed demand for migrant labour in the years to follow, the restrictive attitude was maintained.

The 1970s' economic crisis did not only affect the so-called developed societies. Its causes and effects have also been seen as part of broader processes of economic and political change that have had a global reach. It is precisely in these processes that researchers have found the roots of globalization and, regarding the international flux of people, the beginning of the "age of migrations" (Castles & Miller, 1993). A number of socio-economic transformations have been analysed from different perspectives: the economic restructuring of peripheral economies and the imposition of a neoliberal agenda by the FMI (McNevin, 2009; Mezzadra, 2011; Sassen, 1998; Schierup et al., 2006); the geopolitical shift after the end of the Cold

War and the fall of exit barriers in most countries (Massey, 1999; McNevin, 2009); the out-burst of ethno-national conflicts (Zolberg, 2006); the emergence of transnationalism (Glick-Schiller, Basch, & Blanc-Szanton, 1992); the flexibilization, delocalization and internationalization of productive processes (Schierup et al., 2006); and the development of transportation and communication technologies (Castles & Miller, 1993). As a matter of fact, one of the most significant effects of these complex transformations was the great incentive towards international migrations.

The combination of restrictive policies and sustained demand for labour on the one hand, and of a potentially unlimited supply of migrants on the other, set the conditions for what has been called the "migration crisis" of the 1990s (Castles, 2004). Since these two forces could not match by using the legal channels established by states, alternative strategies quickly developed. Irregular crossing of borders, visa overstaying, the improper use of asylum policy, just to name the most important, became widely used channels to circumvent the states' barriers. Thus, irregular migration emerged as a structural characteristic of current migration processes.

This "unexpected" outcome sharply increased concerns in receiving societies, paving the way for widespread social attention, the anxiety of public opinion and a rapid politicization (Castles, 2004; Vollmer, 2011; Zolberg, 2006). Governments reacted by prioritizing migration policies in their agendas and the main target was precisely irregularity. The result was a multiplication of policies, mechanisms, and investments both at national and international levels, in an attempt to regain control over migration. These extraordinary efforts, nevertheless, have been largely wiped out by counterstrategies enacted by migrants and by those interested in the continuation of the fluxes. These dynamics between states and migrants have been compared to an arms race in which action provokes reaction (Broeders & Engbersen, 2007). The most notable effect of these dynamics has been the diversification of the characteristics and modes of irregularity (Bloch & Chimienti, 2011).

2.3 Numbers

One of the most difficult tasks regarding the study of irregular migration has been assessing the magnitude of the phenomenon. It is precisely its irregular character that provides the reasons for such difficulties. If a certain elusiveness of their objects is an inevitable problem for social sciences, this issue becomes even more complex when the object in question is defined as "irregular". Contrary to what occurs with the majority of other social phenomena, with irregular migration it is not possible to count on official statistics. The ways in and out of irregularity are many and available data are limited to only a part of these fluxes.

At the same time, the politicization and social anxieties that have surrounded irregular migration have been a powerful reason for both administrations and public opinion to ask for numbers. After all, in order to assess the extent of a threat, it is firstly necessary to know its magnitude. This has implied the proliferation of analyses, journalistic reports, and official and unofficial estimations. The sensitive aspect of the topic, especially for states that, on the basis of those numbers, could be publically judged as either efficient or inadequate, entailed an inevitable tendency to manipulation (Dal Lago, 2004; Vollmer, 2011). Numbers have often been exaggerated, minimized, hidden or dramatized, depending on the political goal behind their use. In this respect, Vollmer has underlined the relevance of "number games" in the construction of political discourses about irregular migration (Vollmer, 2011).

The complexities related to the estimations of irregular migration and to their use have raised an interesting scientific debate as to their utility. Some scholars have argued against the proliferation of statistics, by stressing the methodological pitfalls and the political misuse of numbers (Koser, 2010; Triandafyllidou, 2012). Others, on the contrary, have emphasized the necessity for the collection of valuable data (Düvell, 2011b; Koser, 2010). More specific debates have developed around the methodological (Espenshade, 1995; Jandl, 2011; Koser, 2010; Triandafyllidou, 2009) and ethical problems related to the use of statistics (Düvell, Triandafyllidou, & Vollmer, 2010; Triandafyllidou, 2009).

Kraler and Reichel have recently stressed that irregular migration estimations and numeric analysis "can be useful for assessing broad trends regarding the dynamics, patterns, as well as structure of irregular migration" (Kraler & Reichel, 2011, p. 121). While it is true that precise numbers are not attainable and that their use is permanently at risk of political mistreatment, the recent proliferation of estimations of the irregular population in different national contexts is certainly welcomed. For a discussion on general trends and the numerical relevance of irregular migration see, for instance: (Castles, 2010; Düvell, 2011c; Kraler & Reichel, 2011; Triandafyllidou, 2009; Vogel, Kovacheva, & Prescott, 2011). For specific reports by country, see: (Passel, Cohn, & Gonzalez-Barrera, 2013; Triandafyllidou, 2009).

2.4 State Policies and Irregular Migration

As a by-product of the interaction between states and migrations, one important strand of research has focused on the study of those policies that directly or indirectly affect irregular migration. The attention to policies is relatively recent and can be linked to the inability of scholars to fully explain the migration crisis at the end of the twentieth century, using their classic tools. The combination of push-pull theories, agent's microeconomic theories and network theories had been fairly successful in describing migration mechanisms, at least as they occurred in accordance to state will. After the oil crisis of the 1970s, and increasingly throughout the 1980s, theoretical efforts were made to interpret the new setting (Massey et al., 1998). Only in the 1990s, however, did the study of the role of the state become central for migration studies (Massey, 1999; Zolberg, 2000). Since then, a vast and diverse literature has emerged. Most of this work has either implicitly or explicitly dealt with irregular migration. Irregularity, being a sort of nemesis of state policies, has been one of the main targets and somehow the measure of the failure and success of these policies. The consolidation of irregularity as a structural phenomenon in all receiving societies, exactly at a time when major efforts were being made to control migration, raised a number of questions. Were states losing control of their borders and populations? Were there hidden interests that secretly favoured irregular migration? How could policies be improved in order to successfully deter unwanted migrants? In relation to these questions, the study of policies and their evolution appeared as a crucial step in order to understand the opportunity structure within which irregular migration emerges and develops as a social phenomenon.

The research on policies has dealt with four main questions: (a) how and by whom are policies decided?; (b) What are the main types?; (c) How are they implemented? (d) Why do they fail? This chapter will analyse the debate around the first three questions; the fourth will be one of the main topics of Chap. 3.

2.4.1 Policy Formation

A first important issue scholars had to deal with concerned the production of migration policies. Two questions appeared critical: how are policies decided and in which arenas? What actors, forces and interests concur to their configuration? These questions are extremely relevant to the discussion on irregular migration. In order to understand to what extent irregularity is the result of a deliberate policy or not, it becomes crucial to identify what interests have favoured its formulation.

Regarding the relevant actors and ideas, a variety of hypotheses have been proposed. The discussion has generally followed general sociological- and political-science theories on policy formation. Some scholars have emphasized the role of domestic political factors, such as: national identities and cultures (Düvell, 2011b; Freeman, 1995; Jordan, Stråth, & Triandafyllidou, 2003), conception of citizenship (Brubaker, 1992), and migratory history (Arango, 2003; Massey et al., 1998; Zolberg, 2006). Others have focused on the role of domestic actors, for instance: employers, labour unions, interest groups, courts, ethnic groups, trade unions, law and order bureaucracies, police and security agencies, local actors and street-level bureaucrats, and private actors (Abella, 2004; Freeman, 1995; Lahav & Guiraudon, 2006; Piore, 1980; Portes, 1978). In this respect, Czaika and de Haas have stressed that, since migration policy is typically the result of a compromise between multiple potentially-competing interests, it can be useful to pay attention to the "discursive coalitions" that may form (Czaika & De Haas, 2013). Another important branch of research has underlined the relevance of legal frameworks, political institutions and their functioning in establishing the procedures and limits of the bargaining around migration policy (Freeman, 1995; Hollifield, 1992; Joppke, 1998; Lahav & Guiraudon, 2006; Money, 1999; Shughart, William, Tollison, & Kimenyi, 1986). Another has focused on the interests of states as sovereign and self-preserving actors (Rudolph, 2005). Finally, many scholars have focused their attention on the role of forces external to states. Within this line of enquiry, what has been emphasized is the role of the global economy (Sassen, 1998; Wallerstein, 2004), of human rights regimes (Jacobson, 1996; Soysal, 1994), and of international legal frameworks and institutions (for instance the EU) (Geddes, 2001, 2003). For a more detailed analysis of these traditions, a number of review essays on immigration policies provide a wide analytical panorama of them (Meyers, 2000; Money, 2010).

Once the existence of a variety of actors and of frequently-irreconcilable interests has been recognized, attention has shifted to the decision process. In relation to this, different positions have emerged on the role of the state. Garcés-Mascareñas has highlighted two main perspectives: theories that consider states mainly as brokers of civil-society demands (Freeman, 1995) and theories that consider states and their interests as the main force behind migration decisions (Garcés-Mascareñas, 2012; Rudolph, 2005). Within this debate, attempts to produce more complex accounts of state imperatives and functioning have been advanced (Boswell, 2007; Lahav & Guiraudon, 2006; Sciortino, 2000). These efforts will be discussed in detail in Chap. 4.

2.4.2 Policies that Affect Irregular Migration

The study of the policies that affect irregular migration has gone hand in hand with their development. After the oil crisis of 1973, most receiving countries observed a proliferation of policies, mechanisms, administrative structures, and legal frameworks dedicated to dealing with the control of international migrations. The real or perceived sense of failure signalled by the migration crisis of the 1990s intensified the development and implementation of newer and increasingly-sophisticated policies. This perpetual escalation of control measures, on the one hand, and migrants' countermeasures on the other, is far from being concluded in our days. The main consequence for research has been a corresponding proliferation of studies, taxonomies and classifications in the attempt to analyse a constantly evolving landscape. The remainder of this section proposes a classification and a brief description of the main policies that affect irregular migration. It is important to mention that, although the hypothesis of an on-going convergence among state practices has been advanced (Cornelius, Martin, & Hollifield, 1994; Doomernik & Jandl, 2008), national approaches still present important differences. Therefore, each state displays a different combination of policies and a peculiar trend of implementation (Castles & Miller, 1993; de Haas, Natter, & Vezzoli, 2018; Düvell, 2011b; Freeman, 2006; Lahav & Guiraudon, 2006).

A first important distinction in classifying migration policies is the one proposed by Hammar (1985) between immigration policy and immigrant policy (Hammar, 1985). Immigration policies include those directed at controlling and selecting or deterring migration fluxes. Within this broad group, two main subgroups can be distinguished: external control policies and internal control policies (Brochmann & Hammar, 1999; Broeders & Engbersen, 2007; Cornelius, 2005; Cornelius et al., 1994; Doomernik & Jandl, 2008; Van Meeteren, 2010). The first group includes: border enforcement policies (Cornelius & Salehyan, 2007); remote control policies, such as carrier sanctions, international and bilateral agreements, visa regimes and entry policies (Finotelli, 2009; Finotelli & Sciortino, 2013; Garcés-Mascareñas, 2012; Guiraudon & Joppke, 2001; Massey, Durand, & Pren, 2015; Triandafyllidou, 2009; Triandafyllidou & Ambrosini, 2011; Zolberg, 2000, 2006); and policies aimed at reducing push factors in sending countries (for instance, funds for development) (Hollifield, 2004). The second group includes three main sub-groups: (a) policies directed at making irregular residence difficult and costly through labour market controls, for example, employer sanctions, employers' deputation to check for identities, labour site inspections) (Brochmann & Hammar, 1999; Broeders, 2009; Broeders & Engbersen, 2007; Cornelius, 2005) and policies aimed at the exclusion of irregular migrants from public services (identification checks in order to use services) (Broeders, 2009; Van Der Leun, 2003; Van Meeteren, 2010). (b) Policies directed towards the identification, detention and expulsion of irregular migrants (identification and surveillance systems, random checks in public spaces, administrative detention, readmission agreements) (Broeders, 2009; Engbersen & Broeders, 2009; Schinkel, 2009; Schrover et al., 2008; Van Meeteren, 2010). (c) Policies directed at the regularization of irregular migrants (collective and individual regularization, de jure and de facto regularizations) (Baldwin-Edwards & Kraler, 2009; Boswell & D'Amato, 2012; Chauvin, Garcés-Mascareñas, & Kraler, 2013; Engbersen & Broeders, 2009; Finotelli, 2006; Papademetriou, 2005; Schrover et al., 2008).

In a different way, immigrant policies address the management of the immigrant populations, their integration, and the improvement of their living standards (Van Der Leun, 2003). Though usually not explicitly directed towards irregular migrants, these policies can have a tremendous impact on their lives. A first important policy within this group is the one that establishes the limits, rights, conditions and progression of migrants' status towards obtaining citizenship (Chavez, 2007; Finotelli, La Barbera, & Echeverría, 2018; Garcés-Mascareñas, 2012; Isin, 2009; Joppke, 2010; Mezzadra, 2011; Ngai, 2014; Ong, 2005). While the classic distinction among citizens, denizens and aliens (Hammar, 1990) is fundamental, many scholars have shown that a greater variety of statuses and, hence, of hierarchies subsist within those categories (Broeders, 2009; Castles, 2004; Cvajner & Sciortino, 2010b; Finotelli & Sciortino, 2013; López Sala, 2005; Sciortino, 2013; Vasta, 2011). Probably the most relevant aspect of this policy concerns the establishment of the conditions for denizens to keep a regular status and the period of time before eventually becoming citizens. While an open, limitedly-conditioned policy may lead to an efficient progression along statuses, a closed, strongly-conditioned policy may imply drawbacks, slow advance and the possibility of cases of befallen irregularity. A policy within this cluster, that has a direct influence on irregular migrants, is the one that establishes the rights to which they are entitled. In this respect, a variety of arrangements can be discerned, ranging from the absolute exclusion and negation of rights in the Gulf Countries to the full entitlement to social services in countries, like Spain (Arango, 2005; Massey, 1999). For a schematic view of all the main policies affecting irregular migrants, see Table 2.1.

Table 2.1 Policies that directly affect irregular migration

IMMIGRATION POLICIES	**External Control**	Border enforcement	*Border patrolling*
			Surveillance-technology implementation
			Construction of barriers
		Remote control policies	*Carrier sanctions*
			International and bilateral agreements
			Visa regime and entry policies
			Asylum and Refugee Policy
		Policies to reduce push factors	*International cooperation to reduce emigration*
	Internal Control	Dissuasion policies	*Labour-market controls*
			Employers' sanctions
			Employers' deputation to check documents
			Exclusion from social services
			Assisted return policies
		Identification, detention and expulsion	*Identification systems*
			Random checks in public spaces
			Administrative detention
			Expulsion
			Readmission agreements
		Regularization Policies	*Individual regularization*
			Collective regularizations
			De jure regularizations
			De facto regularizations
IMMIGRANT POLICIES	**Residence and citizenship policies**	*Permit conditions, requisites and time-length*	
		Permit renewal condition and requisites	
		Conditions, requisites and timing to acquire citizenship	
	Migrants' rights	*Migrants' entitlements and rights*	
		Irregular migrants' entitlements and rights	

2.4.3 *Policy Implementation*

The efforts to identify, classify and comparatively analyse migration policies in order to understand more fully irregular migration, have proved inadequate. Since the early 1990s, many scholars have highlighted the existence of a gap between the laws and policies stated on paper and what they effectively achieved in "reality" (Cornelius et al., 1994). This awareness stimulated an intense debate over the need for a more comprehensive understanding of policies and their interaction with social life. Within this debate, a group of scholars underlined the necessity to shift the focus from policy formation or policy classification to policy implementation (Castles, 2004; Guiraudon & Lahav, 2000; Van Der Leun, 2003). Whereas many studies existed on laws, explicit regulation, policy documents and decision-making processes, scarce attention had been given to their implementation as well as to the resilience of lower-level counterforces (Lahav & Guiraudon, 2006; Van Der Leun, 2003). As pointed by Van del Leun, a large body of literature not directly concerned with the study of migrations, had already "warned against straightforward ideas about the process of implementation of public policies" (Van Der Leun, 2003, p. 28).

The shift of attention to implementation dramatically increased the complexity of the picture. This has posed a number of methodological and epistemological problems. As long as the focus was on laws and regulations, researchers could refer to the official documents and statements by politicians and administrators. Enquiring into implementation, instead, forced them to get out of the libraries and adopt qualitative strategies to find and recompose the pieces of the puzzle. The information gathered in interviews with politicians, bureaucrats, migrants and many other social actors offered a prism of different perspectives that were rarely coincident. Moreover, because of the sensitive character of the information requested, the probability of getting embellished answers or no answers at all was high.

Notwithstanding these difficulties and the relatively recent attention given to implementation, the efforts made in the last two decades have produced significant results.

On the one hand, theoretical attempts have been made to develop frameworks of analysis. Since every national context produces distinctive practices of implementation, two questions have been raised: (a) what determines the specific mode of implementation? (b) How is it possible to explain differences? Four aspects have been suggested as crucial in order to understand different practices: the peculiar national regulatory styles and traditions; the organizational culture of bureaucracies and the degree of discretionality; the grade of isolation of bureaucracies from external pressures; the social attitude and toleration towards informality (Guiraudon & Lahav, 2000; Jordan et al., 2003; Lahav & Guiraudon, 2006; Van Der Leun, 2003). This approach also implied an extension of the actors to be taken in consideration: not only politicians and legislators, but also bureaucrats, policemen, civil servants, teachers, healthcare servants, etc. The focus had to be given to those "street level bureaucrats" that, at the end of the command chain, really enforce policies (Heyman & Smart, 1999; Scott, 1998).

On the other hand, researchers have analysed policy implementation in different countries with the purpose of detecting possible common trends. Lahav and Guiraudon have indicated an on-going shift of focus in the implementation of policies. While before the migration crisis of the 1990s, controls were limited to border enforcement and were implemented exclusively by states' central institutions, after that, controls have been moving "away from the border and outside of the state" (Guiraudon & Lahav, 2000). This process has followed a threefold strategy: a shift outwards, with the adoption of remote control policies; a shift upwards, with the development of international frameworks for control; a shift downwards, with the delegation of control duties to the local institutional level. In 2008, Doomenrik and Jandl proposed another interpretation of this process. They suggested that states' controls are expanding: forwards, externalizing implementation outside the borders; backwards, adopting internal controls and checks in public places and workplaces; and inwards, with an expansion of the requirements placed on migrants (Doomernik & Jandl, 2008).

Another group of scholars have observed a slow but constant shift in the logic of policy implementation (Broeders, 2009; Broeders & Engbersen, 2007; Engbersen, 2001). Broeders characterized this shift as the alternation between two contradictory logics of exclusion: exclusion from documentation/registration and exclusion through documentation/registration (Broeders, 2009). The first logic intended to exclude irregular migrants, denying them the possibility to acquire the documents necessary to access public services. While it may have been effective in fencing migrants' access to welfare, this logic did not prevent the growth of irregular migration and was ineffective in expelling them. The main objective of the second logic was precisely to make expulsions effective. The correct identification of migrants was the main condition that origin states asked for, in order to accept their citizens back, once they were expelled. While this process has occurred principally in Northern European countries, the second logic has been central to the European Union common policy and seems to be gaining importance in the rest of receiving counties. More in general, many authors have underlined the growing importance for the implementation of migration policies of identification technologies and surveillance systems (Engbersen & Broeders, 2009; Leerkes, 2009).

Finally, a number of scholars have suggested the need to look beyond policies closely related to immigration control in order to fully grasp migration management (Finotelli, 2009; Garcés-Mascareñas, 2012). In a recent article, Czaika and de Haas have written:

> Many policies affect migration such as labour market, macro-economic, welfare, trade and foreign policies. Because they affect fundamental economic migration drivers, their influence might actually be larger than specific migration policies, which perhaps have a greater effect on the specific patterns and selection of migrants rather than on overall magnitude and long-term trends, which seem to be more driven by structural political and economic factors in origin and destination countries (Czaika & De Haas, 2013, p. 5).

It seems possible to conclude that only the joint analysis of the interaction and the implementation of migration and refugee policies, labour market policies and welfare policies allows for a full picture of the framework within which irregular migration emerges and evolves.

2.5 Irregular Migrants Lived Experience

From a very different standpoint, a whole bunch of studies on irregular migration has devoted its attention to enquiring into migrants' lived experiences. These studies, more from a sociological and anthropological perspective, have researched into a number of different issues, producing a vast and differentiated literature. The emphasis was on the agency of migrants, on their interaction with the structures of the receiving society and on the consequences of such interaction on their lives. The leading questions have been the following: (a) How do irregular migrants manage to live in a society that does not recognize them as legitimate members? (b) What strategies do they implement to be able to work? (c) How do public opinion and civil societies react in hosting countries?

For a more schematic analysis of this literature, it was chosen to consider three main broad thematic groups.

2.5.1 *Life, Adaptation and Social Interactions*

The shift of focus from policies to migrants' experiences and social interactions raised important methodological questions and produced a number of different perspectives. Different analytical tools have been proposed to make sense of a complex and dynamic phenomenon in which both structures and individuals' agency need to be considered. The concept of strategy is the one that has been prevalently used in the literature (Engbersen, 2001). Van Nieuwenhuyze has recently used the concept of trajectory. Her aim was "to gain an insight into the transitions and choices made by immigrants, and to explore their decisions and motivations within a specific economic and political opportunity structure"(Van Nieuwenhuyze, 2009, p. 19). Cvajner and Sciortino adopted the concept of career, intended as "a sequence of steps, marked by events defined as significant within the structure of actors' narratives and publicly recognized as such by various audiences"(Cvajner & Sciortino, 2010a, p. 2).

The different emphasis given to either structures or agency has fostered an interesting and on-going discussion on the appropriate understanding of irregular migrants' conditions. Should they be considered as victims that passively undergo the consequences of an unfair destiny or as active agents that consciously choose irregularity as a life strategy (Bloch & Chimienti, 2011)? Are they "modern-day slaves" or "villains" that break the law for their own interests (Anderson & Ruhs, 2010)?

The accounts that have adopted the first perspective have underscored the difficulties experienced by irregular migrants. On the one hand, many scholars have researched on their working and social conditions. A propensity towards precarious work, social immobility, poor housing and limited access to healthcare has been widely registered (Ambrosini, 2011, 2012, 2016, 2018; Bloch & McKay, 2017; Bloch, Sigona, & Zetter, 2009; Chavez, 1991; Goldring & Landolt, 2011; Mahler, 1995; Van Der Leun, 2003; Van Nieuwenhuyze, 2009). Studies on the US case have reached milder conclusions (Chavez, 1991, 1994; Massey & Espinosa, 1997). The extensive analysis of the Dutch case has led Engbersen and his colleagues to propose the marginalization thesis. The main idea is that the enforcement of internal control policies and the augmented pressure on irregular migrants have increasingly deteriorated their social conditions. The impossibility to access social services and get employed have pushed them "further underground", forcing them to accept exploitative conditions or even to turn to crime (Engbersen, Van Der Leun, & Leerkes, 2004; Leerkes, Van Der Leun, & Engbersen, 2012).

On the other hand, the personal feelings, attitudes and identity negotiations that irregular migrants develop in relation to their status have been investigated (Coutin, 2005a; De Genova, 2002; Engbersen, 2001; Fernández-Esquer, Agoff, & Leal, 2017; Vasta, 2011; Willen, 2007). Engbersen has argued that the illegal status is a master status, "a dominant social characteristic overshadowing all other personal characteristics" (Engbersen, 2001, p. 240). In this sense, illegal status influences the establishment of all social relations and migrants need to learn to live as irregular migrants. De Genova underlined how the "palpable sense of deportability" and not deportation itself, has a concrete effect on the existence of migrants. "The spatialized condition of "illegality" reproduces the physical borders of nation-states in the everyday life of innumerable places throughout the interiors of the migrant-receiving states" (De Genova, 2002, p. 439). This way, "a spatialized and typically racialized social condition", that becomes functional to the exploitation of migrants, is produced. Willen has studied how the condition of irregularity and the permanent possibility of being detected translate into observable behaviours and "somatic modes attention" on the part of irregular migrants. "Migrant illegality affect not only the external structure of migrants' worlds, but can also extend their reach quite literally into illegal migrants' "in-ward parts" by profoundly shaping their subjective experiences of time, space, embodiment, sociality, and self" (Willen, 2007, p. 10).

After a critical review of this literature, Van Meeteren has argued that the survival perspective has become a widespread convention. The main limit has been an excessive emphasis on structure over agency and, therefore, a limited ability to acknowledge phenomena like irregular migrants' upward mobility; the inability to distinguish different irregular trajectories and outcomes; a tendency to underestimate the role of migrants' strategies, aspirations and skills. Building on this critique and trying to understand more in depth the incorporation of irregular migrants, Van Meeteren developed a model of analysis based on irregular migrants' aspirations. From this perspective, contexts do not mechanically constrain or construct irregular migrants' actions. Instead, they create a certain window of opportunities and migrants, on the basis of their own personal characteristics, may take advantage and react to it. This implies that, within the same structural context, irregular migrants with different aspirations may attain different grades of incorporation (Van Meeteren, 2010, pp. 31–32).

Although the passive perspective has unquestionably been preponderant, a noteworthy group of scholars have been adopting a different perspective. The acknowledgment that very few irregular migrants live an underground life and that, on the contrary, they generally live in the midst of host societies, has forced some initial persuasions to be reconsidered (Düvell, 2011b). Analyses moved away from dichotomies, like included/excluded or victims/villains. The focus was placed on migrants' individual characteristics and social skills, in the search for variables that could help or deter their incorporation. In this regard, many factors have been discussed, for instance: the role of networks and ethnic communities (Ambrosini, 2017; Cvajner & Sciortino, 2010b; Mahler, 1995; Triandafyllidou, 2017; Van Meeteren, 2010); the role of social, economic and cultural capital (Bourdieu, 1986); the role of time (Cvajner & Sciortino, 2010b); and the role of transnational networks (Portes, 2003; Van Meeteren, 2010). The multiplication of variables in the framework of analysis inevitably leads to a much more complex scenario regarding outcomes. Not only do different contexts set different windows of opportunities but, within each context, different migrants are more or less capable of seizing those opportunities.

On the basis of these developments, in the last few years, efforts have been made in the direction of a diversified understanding of irregular migration (Cvajner & Sciortino, 2010a; Van Meeteren, 2010; Van Nieuwenhuyze, 2009). In this respect, while the most promising tool to advance in this direction is the development of comparative analyses of irregular migrants, in different contexts the available studies are still limited (Van Meeteren, 2010).

2.5.2 Work and Subsistence

Probably the aspect that has received most attention regarding the lived experience of irregular migrants has been related to their economic integration. Also within this debate, a shift from a survival to a more nuanced perspective has been recorded. A number of issues have been researched. First of all, the employment sectors (Baldwin-Edwards & Kraler, 2009; Düvell, 2011b; Kraler & Rogoz, 2011; Vogel et al., 2011). Even if important geographical and contextual differences exist, irregular migrants are usually employed in similar sectors, in particular: agriculture, construction, textile industry, domestic- and care-work, service sector, and prostitution (Düvell, 2011b).

This particular pattern of employment has been widely analysed in connection with the process of restructuring in the economies of the receiving countries. The work of Piore has been path-breaking in signalling the emergence of dual-labour markets (Piore, 1980): on the one hand, highly-skilled, well-paid, secure jobs; on the other, increasingly precarious, insecure, low-skilled jobs. Whereas the segmentation of labour markets was initially considered a pattern affecting only post-industrial economies, the works of Sassen have convincingly shown that it is a feature affecting most of the global economy (Sassen, 1998). Many other scholars have advanced similar analyses and have proposed different concepts to describe this process: flexibilization, informalization, precarization, etc. (Castles, 2010; Goldring & Landolt, 2011; Kloosterman & Rath, 2002; Sassen, 1998; Schierup et al., 2006). A whole sub-category of studies has focused on the relation between the informal economy and irregular migration (Kraler & Rogoz, 2011; Papademetriou, O'Neil, & Jachimowicz, 2004; Reyneri, 2004; Samers, 2004; Sassen, 1998; Triandafyllidou, 2009).

Another group of scholars have studied the employment strategies of irregular migrants. Many tactics have been discovered: informal employment; self-employment; use of fraudulent papers; and renting of authentic papers (Coutin, 2003; Van Meeteren, 2010). As an answer to the increased controls on the labour market, the recurrence to middlemen and sub-contracting has been a widespread strategy (Broeders & Engbersen, 2007; Massey, 1999; Schierup et al., 2006; Sciortino, 2004a). Engbersen and his colleagues have argued that the fight against informal employment may push irregular migrants to constantly change their sector of employment or even to turn to minor criminal activities as the only option to get an income (Broeders & Engbersen, 2007; Engbersen, 2001; Engbersen & Broeders, 2009; Engbersen et al., 2004; Leerkes et al., 2012).

2.5.3 *Irregular Migrants' Counterstrategies*

As just mentioned, an important strand of research on irregular migration has concentrated on the strategies that migrants develop in order to bypass state controls. From this perspective, migrants are all but passive victims of state action. Indeed, they observe, analyse, share information, develop counterstrategies, and adapt to new conditions (Cornelius & Salehyan, 2007; Schweitzer, 2017; Stavilă, 2015). As noted by Düvell, irregular migrants are often individualist and entrepreneurial, highly responsive to labour-market needs and more mobile than indigenous populations (Düvell, 2006, 2011b). To act like this, they can usually count on extensive networks of friends, relatives, co-nationals and co-ethnics.

Various concepts have been proposed to address this complex web of actions, tactics, informal networks, etc. Scott has proposed the concepts of "weapons of the weak" and "shadow institutions" to acknowledge those everyday forms of resistance that are put in place by the less favoured in contexts of social inequality (Scott, 1998, 2008). A similar idea lies behind Hughes's concept of "bastard institutions" (Hughes, 1994 in Leerkes, 2009). Engbersen has suggested the notion of "residence strategies" to refer to those "strategies aimed at prolonging residence and avoid deportation" (Engbersen, 2001, p. 223). More recently, Bommes and his colleagues have used the concept of "foggy social structures" to highlight those "social structures that emerge from efforts by individuals and organizations to avoid the production of knowledge about their activities by making them either unobservable or indeterminable" (Bommes & Kolb, 2002, p. 5 in Engbersen & Broeders, 2009, p. 868; Bommes & Sciortino, 2011;).

As regards the specific tactics developed by irregular migrants, a diversified picture has been sketched. Engbersen has identified six tactics: mobilization of social capital, bogus marriages, manipulation of identity, operating strategically in the public space, legal action, and crime (Engbersen, 2001). Vasta has focused her attention on the functioning of the paper market. She has shown how irregular migrants engage in a dialectic process with the structures and control mechanisms of receiving societies. Buying, renting, and borrowing someone else's papers are part of a productive process by which they permanently construct and re-construct their subjectivity (Vasta, 2011). Van der Leun, working on the Dutch case, has shown how irregular migrants are able to find and actively exploit the loopholes existing in the legislation and in the implementation of control policies (Van Der Leun, 2003). On the one hand, the complexity of legislation, the different dimension and sectors of application and the existence of various and often-uncoordinated levels of governance determine the presence of legal ambiguities, contradictions and voids. On the other, irregular migrants and their networks, often with the help of lawyers, NGO's and even street-level bureaucrats, successfully learn to take advantage of them (Ambrosini, 2017).

2.6 The Consequences of Irregular Migration

To conclude, an important group of studies has enquired into the effects of irregular migration on receiving societies. These have been analysed from a number of perspectives and have usually given way to heated debates. In particular, three questions have been crucial: (a) What are the effects of irregular migration on the economies of the receiving countries? (b) What are political effects? (c) What are the social effects?

2.6.1 Economic Consequences

From an economic point of view, many questions have been raised, for instance, the effects of irregular migration on: production, consumption, fiscal outcome, wages distribution, segmentation of the markets, etc. (Düvell, 2006; Espenshade, 1995; Hanson, 2007; Koser, 2010; Portes, 1978). As pointed out by Hanson, in receiving societies, there is a widespread belief that irregular migration negatively affects the economy. Nevertheless, these ideas are rarely rooted in comprehensive economic analyses and derive more often from politicized opinions or simple prejudices. A more objective approach needs to acknowledge both the benefits (the increased availability of workers, the better use of resources, the boost on tax revenues) and costs (the use of public services and infrastructures, the lowering effect on some wages) of irregularity. Moreover, it needs to consider that these effects are not uniformly distributed and that, while some parts of society may benefit, some others may lose. On the whole, Hanson concludes that irregular migration has a limited impact. In the case where it persists, it is because a strong economic rationale subsists, at least on the part of the productive structure. In particular, for those businesses that are subject to market fluctuations, irregular migration represents a much more efficient and flexible solution than legal migration (Hanson, 2007).

Another well-established idea about irregular migration hypothesizes a substitution effect between irregular migrants and native workers. Research has shown little evidence of this. On the contrary, a complementarity role has appeared more plausible (Düvell, 2011a, 2011b; Jordan & Düvell, 2002; Reyneri, 1998, 2004; Samers, 2004; Van Meeteren, 2010). As pointed out by Düvell, irregular migration may even create "new markets for jobs and allow indigenous populations to enter the labour market" (Düvell, 2011a, p. 64; Young, 1999). In this respect, he presents an example of how the availability of irregular workers can generate a positive economic cycle. Their low wages make it affordable for lower-income households to hire migrants as domestic workers. This, on the one hand, creates a new employment market. On the other, it "frees indigenous women from housework and allows them to re-enter the labour market". Households' incomes increase, state revenues rise and a new market of lower-priced goods and services is generated for low-wage workers (Düvell, 2011a, p. 64).

Considerable attention has been focused on the impact of irregular migration on the welfare state (Baldwin-Edwards, 2004; Bommes & Geddes, 2000; Düvell, 2006, 2011a; Sciortino, 2004b) and, more in general, on the state budget. Bommes and Geddes have underlined that, since every national context is different, generalizations are problematic. Each state is based on a different historically-established concept of nation, a different mode of defining loyalty, a different immigration history, and a specific welfare regime. Each state, then, provides a distinct repertoire of public services by using different organizational infrastructures (Bommes & Geddes, 2000; Esping-Andersen, 1990). This implies that the impact of irregular migration will be necessarily differentiated and that each case needs to be analysed autonomously. On the whole, however, as highlighted by Düvell, "in many countries irregular immigrants have no, or only limited, access to public services and avoid any interaction with statutory agencies; therefore, often there is almost no negative welfare aspect" (Düvell, 2006, 2011a, p. 64).

A number of other possible negative effects of irregular migration have been alleged: unfair labour competition, decrease in wages, displacing of indigenous workers, undermining of power relations between organised workers and employers, tax evasion, illegitimate claim for, or use of, social services, congestion of the housing market, undermining of the rule of law, and exploitation and emergence of criminal milieus. Nevertheless, these phenomena tend to occur on a small scale because numbers are very limited (Düvell, 2011a; Koser, 2010).

2.6.2 Political Consequences and Social Consequences

Political and social consequences of irregular migration are another topic that has been extensively enquired. Also in this case, research has had to struggle against widespread preconceptions.

The idea of an on-going invasion, often fostered by the sensationalized use of images and titles in the media, raised doubts about the actual strength of states. In particular, irregular migration seemed to threaten their sovereignty and endanger their internal security (Broeders, 2009; Dal Lago, 2004; Koser, 2010). This second aspect gained relevance especially after the terrorist attacks in the early 2000s in the US and Europe (Huysmans, 2006). In a number of countries, right-wing parties emerged to mobilize and give voice to anti-immigrant opinions (Freeman, 1995). More in general, a phenomenon that had been until then essentially marginal, started to gain more and more attention and to become the object of public discourses (Kraler & Rogoz, 2011).

Notwithstanding the real extent to which irregular migration challenged receiving states (for a thorough discussion, see Chap. 3), the attention that the phenomenon reached in the public opinion and the fast politicization that followed, induced most governments to give top priority to the issue. The main result, as mentioned before, was a general trend towards restriction and a widespread implementation of policies and initiatives explicitly directed against irregular migrants. The change of paradigm was skilfully represented by the metaphors and slogans that were used: "zero migration policy", "Fortress Europe", "Panopticon Europe", and "prevention through deterrence" (Broeders, 2009; Cornelius & Salehyan, 2007; Engbersen, 2001). These developments had a number of consequences. As regards migration, the financial and human costs of crossing the borders dramatically rose; previously circular or seasonal fluxes transformed into permanent settlement and the role of people smugglers and human traffickers increased (Broeders, 2009; Cornelius, 2005).

As shown in this chapter, irregular migration has received increasing attention in recent decades. This has resulted in a wide and diversified literature that has adopted a number of perspectives and has tried to provide answers to a number of questions. The attempt to briefly review this extensive literature was made not with the aim to exhaustively cover all that has been written. The aim, instead, was to offer an overview of the main issues that have arisen and the main approaches that have been adopted to provide possible answers. This overview has deliberately concentrated principally on the descriptive works or on the descriptive parts of the works that have been analysed. In the next chapter, the focus will shift to the theoretical explanations that have been put forward to explain irregular migration.

Bibliography

Abella, M. (2004). The role of recruiters in labor migration. In D. S. Massey & J. E. Taylor (Eds.), *International migration. Prospects and policies in a global market* (pp. 201–211). Oxford, UK: Oxford University Press.

Ambrosini, M. (2011). Undocumented migrants and invisible welfare: Survival practices in the domestic environment. *Migration Letters, 8*(1), 34–42.

Ambrosini, M. (2012). Surviving underground: Irregular migrants, Italian families, invisible welfare. *International Journal of Social Welfare, 21*(4), 361–371. https://doi.org/10.1111/j.1468-2397.2011.00837.x

Ambrosini, M. (2016). From "illegality" to tolerance and beyond: Irregular immigration as a selective and dynamic process. *International Migration, 54*(2), 144–159. https://doi.org/10.1111/imig.12214

Ambrosini, M. (2017). Why irregular migrants arrive and remain: The role of intermediaries. *Journal of Ethnic and Migration Studies, 43*(11), 1813–1830. https://doi.org/10.1080/1369183X.2016.1260442
Ambrosini, M. (2018). *Irregular immigration in Southern Europe*. Cham, The Netherlands: Springer International Publishing. https://doi.org/10.1007/978-3-319-70518-7
Anderson, B. (2006). *Imagined communities: Reflections on the origin and spread of nationalism*. London: Verso Books.
Anderson, B., & Ruhs, M. (2010). Researching illegality and labour migration. *Population, Space and Place, 16*(3), 175–179. https://doi.org/10.1002/psp.594
Arango, J. (2003). La explicación teórica de las migraciones: luz y sombra. *Migración y Desarrollo, 1*(1), 4–22.
Arango, J. (2005). Dificultades y dilemas de la política de inmigración. *Arbor, 181*(713), 17–25.
Baldwin-Edwards, M. (2004). *Immigrants and the welfare state in Europe*.
Baldwin-Edwards, M. (2008). Towards a theory of illegal migration: Historical and structural components. *Third World Quarterly, 29*(7), 1449–1459.
Baldwin-Edwards, M., & Kraler, A. (2009). *REGINE-Regularisations in Europe*. Amsterdam: Amsterdam University Press.
Bloch, A., & Chimienti, M. (2011). Irregular migration in a globalizing world. *Ethnic and Racial Studies, 34*(8), 1271–1285. https://doi.org/10.1080/01419870.2011.560277
Bloch, A. M., & McKay, S. (2017). *Living on the margins: Undocumented migrants in a global city*. Bristol, UK: Policy Press.
Bloch, A., Sigona, N., & Zetter, R. (2009). *"No Right to Dream": The social and economic lives of young undocumented migrants in Britain*. London: Paul Hamlyn Foundation.
Bommes, M., & Geddes, A. (Eds.). (2000). *Immigration and welfare. Challenging the borders of the welfare state*. London/New York: Routledge.
Bommes, M., & Kolb, H. (2002). Foggy social structures in a knowledge-based society–irregular migration, informal economy and the political system. *Unpublished Paper, University of Osnabruck*.
Boswell, C. (2007). Theorizing migration policy: Is there a third way? *International Migration Review, 41*(1), 75–100.
Boswell, C., & D'Amato, G. (Eds.). (2012). *Immigration and social systems: Collected essays of Michael Bommes*. Amsterdam: Amsterdam University Press.
Bourdieu, P. (1986). The forms of capital. In *Handbook of theory and research offor the sociology of education*. Westport, CT: Greenwood Press.
Brochmann, G., & Hammar, T. (1999). *Mechanisms of immigration control: A comparative analysis of European regulation policies*. Oxford, UK/New York: Berg Pub Limited.
Broeders, D. (2009). *Breaking down anonymity digital surveillance on irregular migrants in Germany and the Netherlands*. Rotterdam, The Netherlands: Erasmus Universiteit.
Broeders, D., & Engbersen, G. (2007). The fight against illegal migration: Identification policies and immigrants' counterstrategies. *American Behavioral Scientist, 50*(12), 1592–1609. https://doi.org/10.1177/0002764207302470
Brubaker, R. (1992). *Citizenship and nationhood in France and Germany*. Cambridge, MA: Harvard University Press.
Castles, S. (2004). Why migration policies fail. *Ethnic and Racial Studies, 27*(2), 205–227. https://doi.org/10.1080/0141987042000177306
Castles, S. (2010). Migración irregular: causas, tipos y dimensiones regionales. *Migración y Desarrollo, 8*(15), 49–80.
Castles, S., & Miller, M. J. (1993). *The age of migration. International population movements in the modem world*. New York.
Chauvin, S., Garcés-Mascareñas, B., & Kraler, A. (2013). Working for legality: Employment and migrant regularization in Europe. *International Migration, 51*(6), 118–131. https://doi.org/10.1111/imig.12109

Chavez, L. R. (1991). Outside the imagined community: Undocumented settlers and experiences of incorporation. *American Ethnologist, 18*(2), 257–278.

Chavez, L. R. (1994). The power of the imagined community: The settlement of undocumented Mexicans and Central Americans in the United States. *American Anthropologist, 96*(1), 52–73.

Chavez, L. R. (2007). The condition of illegality. *International Migration, 45*(3), 192–196.

Cornelius, W. A. (2005). Controlling 'Unwanted' immigration: Lessons from the United States, 1993–2004. *Journal of Ethnic and Migration Studies, 31*(4), 775–794. https://doi.org/10.1080/1369183050011001 7

Cornelius, W. A., Martin, P. L., & Hollifield, J. (1994). *Controlling immigration: A global perspective*. Standford, CA: Stanford University Press.

Cornelius, W. A., & Salehyan, I. (2007). Does border enforcement deter unauthorized immigration? The case of Mexican migration to the United States of America. *Regulation & Governance, 1*(2), 139–153. https://doi.org/10.1111/j.1748-5991.2007.00007.x

Coutin, S. B. (2003). *Legalizing moves: Salvadoran immigrants' struggle for US residency*. Ann Arbor, MI: University of Michigan Press.

Coutin, S. B. (2005a). Being en route. *American Anthropologist, 107*(2), 195–206.

Coutin, S. B. (2005b). Contesting criminality: Illegal immigration and the spatialization of legality. *Theoretical Criminology, 9*(1), 5–33. https://doi.org/10.1177/1362480605046658

Cvajner, M., & Sciortino, G. (2010a). A tale of networks and policies: Prolegomena to an analysis of irregular migration careers and their developmental paths. *Population, Space and Place, 16*(3), 213–225.

Cvajner, M., & Sciortino, G. (2010b). Theorizing irregular migration: The control of spatial mobility in differentiated societies. *European Journal of Social Theory, 13*(3), 389–404.

Czaika, M., & De Haas, H. (2013). The effectiveness of immigration policies. *Population and Development Review, 39*(3), 487–508.

Dal Lago, A. (2004). *Non-persone: l'esclusione dei migranti in una società globale*. Milano: Feltrinelli Editore.

De Genova, N. (2002). Migrant "Illegality" and deportability in everyday life. *Annual Review of Anthropology, 31*(1), 419–447.

De Genova, N. (2004). The legal production of Mexican/migrant "illegality". *Latino Studies, 2*(2), 160–185.

de Haas, H., Natter, K., & Vezzoli, S. (2018). Growing restrictiveness or changing selection? The nature and evolution of migration policies. *International Migration Review, 52*, 324–367. https://doi.org/10.1111/imre.12288

Doomernik, J., & Jandl, M. (Eds.). (2008). *Modes of migration regulation and control in Europe*. Amsterdam: Amsterdam University Press.

Düvell, F. (2006). *Illegal immigration in Europe: Beyond control?* Basingstoke, UK/New York: Palgrave Macmillan.

Düvell, F. (2011a). Irregular immigration, economics and politics. *CESifo DICE Report, 9*(3), 60–68.

Düvell, F. (2011b). Paths into irregularity: The legal and political construction of irregular migration. *European Journal of Migration and Law, 13*(3), 275–295. https://doi.org/10.1163/157181611X587856

Düvell, F. (2011c). The pathways in and out of irregular migration in the EU: A comparative analysis. *European Journal of Migration and Law, 13*(3), 245–250. https://doi.org/10.1163/157181611X587838

Düvell, F., Triandafyllidou, A., & Vollmer, B. (2010). Ethical issues in irregular migration research in Europe. *Population, Space and Place, 16*(3), 227–239.

Engbersen, G. (2001). The unanticipated consequences of Panopticon Europe. In V. Guiraudon & C. Joppke (Eds.), *Controlling a new migration world* (pp. 222–246). London, New York: Routledge.

Engbersen, G., & Broeders, D. (2009). The state versus the Alien: Immigration control and strategies of irregular immigrants. *West European Politics, 32*(5), 867–885. https://doi.org/10.1080/01402380903064713

Engbersen, G., & Van Der Leun, J. (2001). The social construction of illegality and criminality. *European Journal on Criminal Policy and Research, 9*(1), 51–70.

Engbersen, G., Van Der Leun, J., & Leerkes, A. (2004). *The Dutch migration regime and the rise in crime among illegal immigrants* (pp. 25–28). Presented at the fourth annual conference of the European Society of Criminology, Global Similarities, Local Differences. Amsterdam, August.

Espenshade, T. J. (1995). Unauthorized immigration to the United States. *Annual Review of Sociology, 21*, 195–216.

Esping-Andersen, G. (1990). *The three worlds of welfare capitalism*. Princeton NJ: Princeton University Press.

Fernández-Esquer, M. E., Agoff, M. C., & Leal, I. M. (2017). Living *Sin Papeles*: Undocumented Latino workers negotiating life in "Illegality". *Hispanic Journal of Behavioral Sciences, 39*(1), 3–18. https://doi.org/10.1177/0739986316679645

Finotelli, C. (2006). La inclusión de los inmigrantes no deseados en Alemania y en Italia: entre acción humanitaria y legitimación económica. *Circunstancia*, (10).

Finotelli, C. (2009). The north–south myth revised: A comparison of the Italian and German migration regimes. *West European Politics, 32*(5), 886–903. https://doi.org/10.1080/01402380903064747

Finotelli, C., La Barbera, M., & Echeverría, G. (2018). Beyond instrumental citizenship: The Spanish and Italian citizenship regimes in times of crisis. *Journal of Ethnic and Migration Studies, 44*(14), 2320–2339. https://doi.org/10.1080/1369183X.2017.1345838

Finotelli, C., & Sciortino, G. (2013). Through the gates of the fortress: European visa policies and the limits of immigration control. *Perspectives on European Politics and Society, 14*(1), 80–101.

Freeman, G. P. (1995). Modes of immigration politics in liberal democratic states. *International Migration Review, 29*(4), 881–902.

Freeman, G. P. (2006). National models, policy types, and the politics of immigration in liberal democracies. *West European Politics, 29*(2), 227–247. https://doi.org/10.1080/01402380500512585

Garcés-Mascareñas, B. (2012). *Labour migration in Malaysia and Spain: Markets, citizenship and rights*. Amsterdam: Amsterdam University Press.

Geddes, A. (2001). Immigration and European integration: Towards fortress Europe? *Refugee Survey Quarterly, 20*(1), 229.

Geddes, A. (2003). *The politics of migration and immigration in Europe*. London: Sage.

Glick-Schiller, N., Basch, L., & Blanc-Szanton, C. (1992). Towards a definition of transnationalism. *Annals of the New York Academy of Sciences, 645*(1), ix–xiv.

Goldring, L., & Landolt, P. (2011). Caught in the work–citizenship matrix: The lasting effects of precarious legal status on work for Toronto immigrants. *Globalizations, 8*(3), 325–341. https://doi.org/10.1080/14747731.2011.576850

Guiraudon, V., & Joppke, C. (Eds.). (2001). *Controlling a new migration world* (Vol. 4). London/New York: Routledge.

Guiraudon, V., & Lahav, G. (2000). A reappraisal of the state sovereignty debate: The case of migration control. *Comparative Political Studies, 33*(2), 163–195. https://doi.org/10.1177/0010414000033002001

Haidinger, B. (2007). *Migration and irregular work in Europe. Undocumented worker transitions*.

Hammar, T. (1985). *European immigration policy: A comparative study*. Cambridge, UK: Cambridge University Press.

Hammar, T. (1990). *Democracy and the nation state*. Aldershot, UK: Avebury.

Hanson, G. H. (2007). *The economic logic of illegal immigration*. New York: Council on Foreign Relations.

Heyman, J., & Smart, A. (1999). States and illegal practices: An overview. In J. Heyman (Ed.), *States and illegal practices*. Oxford, UK: Berg Publishers.

Hobsbawm, E. J. (2012). *Nations and nationalism since 1780: Programme, myth, reality*. Cambridge, UK: Cambridge University Press.

Hollifield, J. (1992). *Immigrants, markets, and states: The political economy of postwar Europe*. Cambridge, MA: Harvard University Press.
Hollifield, J. (2004). The emerging migration state. *International Migration Review, 38*(3), 885–912.
Hughes, E. C. (1994). Bastard institutions. In *Everett C. Hughes on work, race, and the sociological imagination* (pp. 192–199). Chicago: University of Chicago Press.
Huysmans, J. (2006). *The politics of insecurity: Fear, migration and asylum in the EU*. London/New York: Routledge.
İçduygu, A. (2007). The politics of irregular migratory flows in the Mediterranean Basin: Economy, mobility and 'Illegality'. *Mediterranean Politics, 12*(2), 141–161. https://doi.org/10.1080/13629390701373945
Isin, E. F. (2009). Citizenship in flux: The figure of the activist citizen. *Subjectivity, 29*(1), 367–388.
Jacobson, D. (1996). *Rights across borders: Immigration and the decline of citizenship*. Leiden, The Netherlands: Brill.
Jandl, M. (2011). Methods, approaches and data sources for estimating stocks of irregular migrants. *International Migration, 49*(5), 53–77.
Joppke, C. (1998). Why liberal states accept unwanted immigration. *World Politics, 50*(02), 266–293.
Joppke, C. (2010). *Citizenship and immigration*. Cambridge: Polity, UK. Retrieved from http://books.google.es/books?id=5X_Avn3j7j4C
Jordan, B., & Düvell, F. (2002). Irregular migration. *Reading, 100*, 296.
Jordan, B., Stråth, B., & Triandafyllidou, A. (2003). Comparing cultures of discretion. *Journal of Ethnic and Migration Studies, 29*(2), 373–395. https://doi.org/10.1080/1369183032000079648
Kloosterman, R., & Rath, J. (2002). Working on the fringes. Immigrant businesses, economic integration and informal practices. In *Marginalisering Eller Integration* (pp. 177–188). *Stockholm: NUTEK*.
Koser, K. (2010). Dimensions and dynamics of irregular migration. *Population, Space and Place, 16*, 181–193. https://doi.org/10.1002/psp.587
Kraler, A., & Reichel, D. (2011). Measuring irregular migration and population flows – What available data can tell: Measuring irregular migration and population flows. *International Migration, 49*(5), 97–128. https://doi.org/10.1111/j.1468-2435.2011.00699.x
Kraler, A., & Rogoz, M. (2011). *Irregular migration in the European Union since the turn of the millennium–development, economic background and discussion* (Working paper 11/2011).
Lahav, G., & Guiraudon, V. (2006). Actors and venues in immigration control: Closing the gap between political demands and policy outcomes. *West European Politics, 29*(2), 201–223. https://doi.org/10.1080/01402380500512551
Leerkes, A. (2009). *Illegal residence and public safety in the Netherlands*. Amsterdam: Amsterdam University Press.
Leerkes, A., Van Der Leun, J., & Engbersen, G. (2012). *Crime among irregular immigrants and the influence of crimmigration processes* (pp. 267–288). Presented at the Social control and justice: Crimmigration in the Age of Fear.
López Sala, A. M. (2005). La inmigración irregular en la investigación sociológica. In D. Godenau & V. M. Zapata Hernández (Eds.), *La migración irregular una aproximación multidisciplinar* (pp. 161–180). Santa Cruz de Tenerife, Spain: Cabildo Insular de Tenerife, Área de Desarrollo Económico.
Lucassen, J., & Lucassen, L. (2005). *Migration, migration history, history: Old paradigms and new perspectives*. Bern, Switzerland: Peter Lang.
Mahler, S. J. (1995). *American dreaming: Immigrant life on the margins*. Princeton, NJ: Princeton University Press.
Massey, D. S. (1999). International migration at the dawn of the twenty-first century: The role of the state. *Population and Development Review, 25*(2), 303–322.

Massey, D. S., Arango, J., Hugo, G., Kouaouci, A., Pellegrino, A., & Taylor, J. E. (1998). *Worlds in motion. Understanding international migration at the end of the millennium* (pp. 1–59). Oxford, UK/New York: Clarendon Press.

Massey, D. S., Durand, J., & Pren, K. A. (2015). Border enforcement and return migration by documented and undocumented Mexicans. *Journal of Ethnic and Migration Studies, 41*(7), 1015–1040. https://doi.org/10.1080/1369183X.2014.986079

Massey, D. S., & Espinosa, K. E. (1997). What's driving Mexico-US migration? A theoretical, empirical, and policy analysis. *American Journal of Sociology, 102*, 939–999.

McNevin, A. (2009). Contesting citizenship: Irregular migrants and strategic possibilities for political belonging. *New Political Science, 31*(2), 163–181. https://doi.org/10.1080/07393140902872278

Meyers, E. (2000). Theories of international immigration policy-A comparative analysis. *International Migration Review, 34*, 1245–1282.

Mezzadra, S. (2011). The gaze of autonomy. Capitalism, migration and social struggles. In V. Squire (Ed.), *The contested politics of mobility: Borderzones and irregularity* (pp. 121–143). London: Routledge.

Money, J. (1999). *Fences and neighbors. The political geography of immigration control*. Ithaca, NY: Cornell University Press.

Money, J. (2010). Comparative immigration policy. *International studies association compendium project–Sample essays*. Blackwell Publishing. Available at http://www.isacompss.com/info/samples/comparativeimmigrationpolicy_sample.pdf. Accessed on 20 July 2009.

Morokvasic, M. (1993). 'In and out' of the labour market: Immigrant and minority women in Europe. *Journal of Ethnic and Migration Studies, 19*(3), 459–483.

Ngai, M. M. (2014). *Impossible subjects: Illegal aliens and the making of modern America*. Princeton, NJ: Princeton University Press.

Ong, A. (2005). (Re) articulations of citizenship. *Political Science & Politics, 38*(04), 697–699.

Papademetriou, D. (2005). The "Regularization" option in managing illegal migration more effectively: A comparative perspective'. *Migration Policy Brief, September*, (4).

Papademetriou, D., O'Neil, K., & Jachimowicz, M. (2004). *Observations on regularization and the labour market performance of unauthorized and regularized immigrants*. Brussels, Belgium: EC, DG Employment and Social Affairs.

Passel, J. S., Cohn, D., & Gonzalez-Barrera, A. (2013). *Population decline of unauthorized immigrants stalls, may have reversed*. Washington, DC: Pew Hispanic Center. http://www.pewhispanic.org/files/2013/09/unauthorized-sept-2013-FINAL.pdf

Piore, M. J. (1980). *Birds of passage*. Cambridge Books.

Portes, A. (1978). Introduction: Toward a structural analysis of illegal (undocumented) immigration. *International Migration Review, 12*, 469–484.

Portes, A. (2003). Conclusion: Theoretical convergencies and empirical evidence in the study of immigrant transnationalism. *International Migration Review, 37*(3), 874–892.

Reyneri, E. (1998). The role of the underground economy in irregular migration to Italy: Cause or effect? *Journal of Ethnic and Migration Studies, 24*(2), 313–331.

Reyneri, E. (2004). Immigrants in a segmented and often undeclared labour market. *Journal of Modern Italian Studies, 9*(1), 71–93.

Rudolph, C. (2005). Sovereignty and territorial borders in a global age. *International Studies Review, 7*(1), 1–20.

Samers, M. (2004). The 'underground economy', immigration and economic development in the European Union: An agnostic-skeptic perspective. *International Journal of Economic Development, 6*(2), 199–272.

Sassen, S. (1996). *Losing control? Sovereignty in an age of globalization*. New York: Columbia University Press.

Sassen, S. (1998). *Globalization and its discontents: Essays on the new mobility of people and money*. New York: New Press.

Schierup, C.-U., Hansen, P., & Castles, S. (2006). *Migration, citizenship, and the European welfare state: A European dilemma*. OUP Catalogue.

Schinkel, W. (2009). "Illegal Aliens" and the state, or: Bare bodies vs the zombie. *International Sociology, 24*(6), 779–806. https://doi.org/10.1177/0268580909343494

Schrover, M., Van Der Leun, J., Lucassen, L., & Quispel, C. (2008). *Illegal migration and gender in a global and historical perspective*. Amsterdam: Amsterdam University Press.

Schweitzer, R. (2017). Integration against the state: Irregular migrants' agency between deportation and regularisation in the United Kingdom. *Politics, 37*(3), 317–331. https://doi.org/10.1177/0263395716677759

Sciortino, G. (2000). Toward a political sociology of entry policies: Conceptual problems and theoretical proposals. *Journal of Ethnic and Migration Studies, 26*(2), 213–228.

Sciortino, G. (2004a). Between phantoms and necessary evils. Some critical points in the study of irregular migrations to Western Europe. *IMIS-Beiträge, 24*, 17–43.

Sciortino, G. (2004b). Immigration in a Mediterranean welfare state: The Italian experience in comparative perspective. *Journal of Comparative Policy Analysis: Research and Practice, 6*(2), 111–129.

Sciortino, G. (2013). The regulation of undocumented migration. In M. Martiniello & J. Rath (Eds.), *An introduction to international migration studies* (pp. 349–375). Amsterdam: Amsterdam University Press.

Sciortino, G. (2017). *Rebus immigrazione*. Bologna, Italy: Il Mulino.

Scott, J. C. (1998). *Seeing like a state: How certain schemes to improve the human condition have failed.* New Haven, CT: Yale University Press.

Scott, J. C. (2008). *Weapons of the weak: Everyday forms of peasant resistance*. New Haven, CT: Yale University Press.

Shughart, I., William, F., Tollison, R. D., & Kimenyi, M. S. (1986). The political economy of immigration restrictions. *Yale Journal on Regulation, 4*, 79–97.

Soysal, Y. N. (1994). *Limits of citizenship: Migrants and postnational membership in Europe*. Chicago: University of Chicago Press.

Stavilă, A. (2015). No land's man: Irregular migrants' challenge to immigration control and membership policies. *Ethnic and Racial Studies, 38*(6), 911–926. https://doi.org/10.1080/01419870.2014.973431

Torpey, J. (1998). Coming and going: On the state monopolization of the legitimate "means of movement". *Sociological Theory, 16*(3), 239–259.

Torpey, J. (2000). *The invention of the passport: Surveillance, citizenship and the state*. Cambridge, UK: Cambridge University Press.

Triandafyllidou, A. (2009). Clandestino project final report. *Athen, 23*, 70.

Triandafyllidou, A. (2012). *Irregular migration in Europe: Myths and realities*. Ashgate Publishing, Ltd.

Triandafyllidou, A. (2017). Beyond irregular migration governance: Zooming in on migrants' agency. *European Journal of Migration and Law, 19*(1), 1–11. https://doi.org/10.1163/15718166-12342112

Triandafyllidou, A., & Ambrosini, M. (2011). Irregular immigration control in Italy and Greece: Strong fencing and weak gate-keeping serving the labour market. *European Journal of Migration and Law, 13*(3), 251–273.

Van Der Leun, J. (2003). *Looking for loopholes: Processes of incorporation of illegal immigrants in the Netherlands*. Amsterdam: Amsterdam University Press.

Van Meeteren, M. (2010). *Life without papers: Aspirations, incorporation and transnational activities of irregular migrants in the Low Countries*. Rotterdam, The Netherlands: Erasmus Universiteit.

Van Nieuwenhuyze, I. (2009). *Getting by in Europe's urban labour markets: Senegambian migrants' strategies for survival, documentation and mobility*. Amsterdam: Amsterdam University Press.

Vasta, E. (2011). Immigrants and the paper market: Borrowing, renting and buying identities. *Ethnic and Racial Studies, 34*(2), 187–206. https://doi.org/10.1080/01419870.2010.509443

Vogel, D., & Cyrus, N. (2008). *Irregular migration in Europe–Doubts about the effectiveness of control strategies*. Policy Brief, 9.

Vogel, D., Kovacheva, V., & Prescott, H. (2011). The size of the irregular migrant population in the European Union–counting the uncountable? *International Migration, 49*(5), 78–96.

Vollmer, B. A. (2011). Policy discourses on irregular migration in the EU – 'Number Games' and 'Political Games'. *European Journal of Migration and Law, 13*(3), 317–339. https://doi.org/10.1163/157181611X587874

Wallerstein, I. M. (2004). *World-systems analysis: An introduction*. Durham, NC and London: Duke University Press.

Willen, S. S. (2007). Exploring "Illegal" and "Irregular" migrants' lived experiences of law and state power. *International Migration, 45*(3), 2–7.

Williams, C. C., & Windebank, J. (1998). *Informal employment in the advanced economies: Implications for work and welfare*. Psychology Press.

Wimmer, A., & Glick-Schiller, N. (2003). Methodological nationalism, the social sciences, and the study of migration: An essay in historical epistemology. *International Migration Review, 37*(3), 576–610.

Young, B. (1999). *Die Herrin und die Magd: Globalisierung und die neue internationale Arbeitsteilung im Haushalt*. na.

Zolberg, A. (2000). The politics of immigration policy: An externalist perspective. In *Immigration Research for a New Century* (pp. 60–68).

Zolberg, A. (2003). The archaeology of "remote control". In A. Fahrmeir, O. Faron, & P. Weil (Eds.), *Migration control in the North Atlantic World. The evolution of state practices in Europe and the United States from the French Revolution to the inter-war period* (pp. 195–222). New York: Berghahn Books.

Zolberg, A. (2006). Managing a world on the move. *Population and Development Review, 32*(S1), 222–253.

Zolberg, A., & Benda, P. M. (2001). *Global migrants, global refugees: Problems and solutions*. New York: Berghahn Books.

Chapter 3
Irregular Migration Theories

While in Chap. 2 a general and introductory overview of the main research lines on irregular migration was presented, in this chapter the focus will centre on the theoretical accounts that have been proposed to explain irregular migration. The aim is to analyse how the main research question of this book – how can irregular migration be explained? – has been addressed and what have been the main theoretical hypotheses proposed so far. The chapter will be divided into four parts. In the first part, the so-called gap hypothesis and the debate that has surrounded it will be discussed. This debate is particularly relevant for the discussion because the arguments and positions that have emerged in that context have strongly influenced the theoretical treatment of irregular migration. Since irregular migration was one of the main indicators of the existence of a gap between policy goals and outcomes, the explanations for the latter became an immediate way to understand the former. Irregular migration, from this perspective, was interpreted as the result of whether policy failure or policy choice. As the debate evolved, interpretations become more varied and the two phenomena were more clearly distinguished. Nevertheless, the gap logic remained the dominant framework behind most theories of irregular migration. Accordingly, almost all these theories, although in different ways, have followed one of two basic arguments that have been offered to explain the gap hypothesis. In the second part of the chapter, those theories that have followed the first argument, i.e. the idea of irregular migration as the result of states' diminished control capacities will be presented. In the third part, the theories influenced by the second argument, i.e. irregular migration intended as the outcome of states' implicit or explicit choices will be discussed. Finally, in the last part, there will be a critical discussion of the strengths and weaknesses discernible in the current theoretical understanding of irregular migration.

G. Echeverría, *Towards a Systemic Theory of Irregular Migration*, IMISCOE Research Series, https://doi.org/10.1007/978-3-030-40903-6_3

3.1 The Gap Hypothesis Debate

In their 1994 book *The ambivalent quest for immigration control*, Cornelius, Martin and Hollifield (Cornelius, Martin and Hollifield, 1994), after having comparatively analysed the immigration policy and policy outcomes in nine industrialized democracies, proposed two interrelated theses. On the one hand, they suggested a "convergence hypothesis". This stated that a growing similarity was observable among the states they had analysed, in particular concerning: the policy instruments adopted to control immigration; the results of immigration control measures; social integration policies; the public opinion reaction to immigration flows and governments efficacy. On the other hand, they suggested a "gap hypothesis": "the gap between *goals* of national immigration policy (laws, regulation, executives actions, etc.) and the actual results of policies in this areas (policy *outcomes*)", they wrote, "is wide and growing wider in all major industrialized democracies, thus provoking greater public hostility towards immigrants in general (regardless of legal status) and putting intense pressure on political parties and government official to adopt more restrictive policies" (Cornelius et al., 1994, p. 3). Irregular migration, from their perspective, was the result of "the administrative, political and economic difficulties that hinder the enforcement of laws and regulations against it in open and pluralistic societies" (Cornelius et al., 1994, p. 4). These difficulties responded to various factors, but two seemed crucial: the strength of push and pull forces that strongly encouraged migrations, and the rise of rights-based politics that severely limited states' capacities.

The book was not the first to address these issues. Especially in the US, there had already been many contributions on irregular migration and control policies (Bean, Edmonston, & Passel, 1990; Chavez, 1991; Chiswick, 1988; Cornelius, 1982; Espenshade, 1995; Hollifield, 1992; Massey, 1987; Passel, 1986; Piore, 1980; Portes, 1978; Portes & Bach, 1985). However, Cornelius, Martin and Hollifield's work was able to reframe the debate around its theses and to orient much of the debate in the years that followed. As a demonstration of this, there exists a large number of books and articles that have explicitly referred to the gap hypothesis, either contesting it, supporting it or developing it (Castles, 2004; Cornelius & Rosenblum, 2005; Cornelius & Tsuda, 2004; Czaika & de Haas, 2011; Freeman, 1995; Guiraudon & Joppke, 2001; Guiraudon & Lahav, 2000; Joppke, 1998a, 1998c; Lahav & Guiraudon, 2006; Sassen, 1996; Zolberg, 2000). In particular, two issues have animated this debate: firstly, the actual existence and the possible "size" of the gap; secondly, the nature and origin of the gap.

3.1.1 *Is There a Gap?*

Many contradictory positions have emerged regarding this question. A number of scholars have been critical of the very concept of a "gap hypothesis". Joppke, for instance, has argued that the notion of an emergent gap between policy goals and policy outcomes may suggest that there has been a moment in which these two coincided. In particular, regarding migration, it may be that, at a certain point in history, states, on the basis of their absolute sovereign power, had been perfectly able to control the movements of populations. This notion, however, "is premised on a simplistic and static notion of sovereignty, thus denying its historical variability and chronic imperfection" (Joppke, 1998c, p. 267). Building on this critique, Joppke suggested that the gap is an inevitable fact, and that what needs to be hypothesized is not its existence, but rather its magnitude and causes. Since sovereignty has rarely been absolute, the attention should centre on the degree to which states are able to implement rules and on the reasons that strengthen or weaken that capacity.

The bulk of the debate has focused on the real extent to which states may or may not be losing control over migrations (for a review of this debate see: Czaika & de Haas, 2011; Schinkel, 2009). In this respect, two main positions have developed. On the one hand, there are those who believe that states have lost much of their power to control migrations and that policies have become largely ineffective (Castles & Miller, 1993; Cornelius et al., 1994; Cornelius & Tsuda, 2004; Jacobson, 1996; Sassen, 1996, 1998). These positions have resonated with the broader idea, developed by globalization theorists, that states are slowly losing their prerogatives and becoming a "zombie-category" (Schinkel, 2009). On the other hand, there are those who contest this hypotheses and believe, instead, that the power of states and their efficacy have actually increased (Brochmann & Hammar, 1999; Freeman, 1995; Guiraudon & Lahav, 2000; Joppke, 1998c). From this perspective, the gap between goals and outcomes in migration management has not to do with a diminished capacity, but with states' choices or states' self-limitation.

In a recent article, Czaika and de Haas have extensively analysed how this debate has evolved through the 2000s (Czaika & de Haas, 2011). Whereas the two positions had initially been mainly theoretical, as time passed, the arguments have been strengthened on the basis of empirical researches. The increased availability of data and of case studies, however, has not been sufficient to solve the dispute. In fact, the divide has expanded as the results obtained through quantitative analysis (policies are effective) and those obtained through qualitative ones (policies are not effective) have delivered contrasting responses. "How" then "can we explain that various migration policy instruments turn out to be significantly effective, and that, nevertheless, migration policies are often perceived as not reaching their stated and intended objectives?" ask Czaika and de Haas (2011, p. 4). In the authors' opinion, this seemingly unsolvable incongruence has to do with the conceptual confusion and the lack of precision that have generally characterized the theoretical debate. In particular, the authors have underlined three critical aspects. Firstly, they have argued that there has been ambiguity behind the concept of policy effectiveness. Does it refer to, and does it have to be measured in relation to, the desired effect or to the actual effect of policies? Secondly, there has been little attention paid to distinguishing the different time-scales and levels of aggregation within which policies act. "The empirical literature on policy effects generally focuses on the effects of *specific* measures on *specific* (primarily legally defined) categories of migration over relatively limited time periods, the qualitative literature on migration policy effects tends to address the effects of *overall* levels of policy restrictiveness on *overall* (gross) and long-term volumes, trends and patterns of international migration" (Czaika & de Haas, 2011, p. 4). Finally, there has been a problem regarding the difference between what is stated in policy discourses or even in laws and what is effectively implemented.

3.1.2 What Gaps?

The points proposed by Czaika and de Haas, actually resume a line of criticism that emerged after the gap hypothesis was proposed. Many scholars, in fact, departing from marked evidence that policy discourses could not be taken as policy enactments, started to analyse the different dimensions that the gap hypothesis included within its main idea. Not only was it possible to recognize a gap between policy goals and policy outcomes, but one could also be observed between policy discourses and policy implementation. Along this path of enquiry, a number of other gaps have been identified which have been particularly interesting in relation to the interpretation of irregular migration (Cornelius & Tsuda, 2004; De Genova, 2004; Lahav & Guiraudon, 2006). The main gaps that have been identified will be discussed, following the threefold scheme proposed by Czaika and de Haas (Czaika & de Haas, 2011, pp. 18–23), and a fourth gap will be added, which they have not considered.

The first gap is the so-called discursive gap. This gap deals with the distance that is always discernible, in all political contexts, between what is stated in political discourses and what is then actually put into effect in laws, measures and regulations. Accordingly, it would be a mistake to measure policy effectiveness in relation to policy discourses. A much more accountable benchmark for a realistic evaluation would be to consider what is actually written in the executive dispositions. This issue, as many scholars have underlined, has become particularly relevant since the migration crisis of the 1990s. In fact, the widespread anxieties about migration and the strong politicization that followed in many countries determined an escalation of the anti-immigrant rhetoric by both politicians and administrators. While this has certainly implied a change in the discourses and the promise of many and widely-publicized super-restrictive initiatives, a closer analysis of the actual decisions may suggest a milder reality. As a matter of fact, an objective evaluation of policies has become increasingly difficult. Within this context, moreover, various scholars have detected the spread of what has been called "symbolic policies", i.e., policies focused more on publically suggesting severity rather than on actually achieving it (Andreas, 1998; Castles, 2004; De Genova, 2004; Freeman, 1995; Massey et al., 1998). A number of factors have been put forward in relation to the discursive gap: the existence of hidden agendas; the role of populist politics; the diversified social interests; the complexity of the policy-bargaining once television cameras are switched off; the various political, legal, economic domestic and international constraints; the fact that migration discourses are general and migration policies are specific (Castles, 2004; Cornelius & Tsuda, 2004; Czaika & de Haas, 2011).

The second gap is the so-called implementation gap. Here the problem is related to the distance existing between what is written in the papers regarding laws, measures and regulations, and what is actually implemented by the administrations at their various levels. From this perspective, it would be equally misleading to evaluate policy effectiveness in relation to what is stated in the official documents. In fact, a crucial and decisive element regarding migration policies concerns how they are effectively implemented. Also in this case, various causes may determine a greater or smaller implementation gap: the peculiar national regulatory styles and traditions; the organizational culture of bureaucracies and the degree of discretionality; the grade of insulation of bureaucracies from external pressures; possible intra-administration conflicts or scarce coordination; the social attitude towards and toleration of informality; budgetary constraints; corruption (Cornelius & Tsuda, 2004; Czaika & de Haas, 2011; Guiraudon & Lahav, 2000; Jordan, Stråth & Triandafyllidou, 2003; Lahav & Guiraudon, 2006; Van Der Leun, 2003). The implementation gap not only poses conceptual difficulties but also methodological ones. It is self-evident that researching on the daily work of thousands of street-level bureaucrats or quantitatively measuring implementation could prove to be a prohibitive task.

Czaika and De Haas have referred to the third gap as the efficacy gap, meaning: "the extent to which a change in an effectively implemented policy has the capacity to produce an effect" (Czaika & de Haas, 2011, p. 22). The point here is that even a meticulously written, grounded and implemented policy may reach different results from those expected. The measurement of the efficacy gap may vary from complete failure to a very close attainment of the desired effects. The variables that intervene at this level have to do with the fact that policies do not act in a social void; on the contrary, they interact with a complex and dynamic web of actors and forces that have their own goals and strategies. In this regard, a number of possible limitations to policy effectiveness need to be considered: unintended consequences; implementation failure; unexpected interactions with other policies; counterstrategies on the part of migrants. Moreover, a temporal factor needs to be taken into account. Whereas the effects of a policy may appear satisfactory in the short run, in the medium, long run they could become ineffective or even counterproductive. With respect to this, Freeman (1995) has explicitly talked about the "temporal illusion" of migration policy: "the effects of migration tend to be lagged; the short-term benefits oversold and the long-term costs denied or hidden to show up clearly only in the outyears" (Freeman, 1995).

A fourth gap, very much related to the third, could be referred to as the knowledge or epistemological gap. This gap is concerned with the limits inherent to all processes of knowledge production that are the necessary preceding step for policy design and implementation (Bommes & Kolb, 2002; Engbersen & Broeders, 2009; Freeman, 1995; Scott, 1998). The simplest example of this gap may be found in the impossibility to precisely count irregular migrants. How could a policy be effective if it is directed towards a phenomenon that is not even possible to quantify? Yet, limiting to the counting problem risks understating the magnitude of the issue. In fact, the problem lies not only in having to deal with the impossibility of using statistical tools or producing rigorous numerical figures, but it also lies in the complexity of social interactions and the impossibility of producing accurate, all-embracing descriptions of it. The "illusion of control", that the discussed gaps have evidenced, has perhaps primarily to do with the "illusion of knowledge". The knowledge gap calls attention to this point: every perspective is a partial, imperfect and inevitably biased viewpoint on reality. It, therefore, affects those who deliver policy discourses, those who write laws, regulations and measures, those who produce white papers, those who implement policies and, of course, those who study the effects of those policies.

Considering the logic behind the four types of gaps, it seems possible to clearly distinguish two main explanations. The efficacy and knowledge gap explains the mismatch between policy goals and outcomes as the result of state failure, despite its efforts. The discursive and implementation gap, on the contrary, suggests a certain degree of complicity on the part of the state, and the mismatch as a somewhat intentional outcome. Although the theories advanced to explain irregular migration have offered a great variety of explanations, they all seem to generally follow one of these two rationalities. For this reason, as the attention will now shift to these theories, two main groups will be distinguished (Table 3.1).

3.2 Irregular Migration as States' Failure

The focus of the discussion will now move to the theories and hypotheses that, implicitly or explicitly, have proposed an explanation for irregular migration. Rather than presenting the different approaches following the theses of single scholars, disciplinary distinctions, or chronological accounts, the choice has been to try to identify the main, broad explanatory lines that have emerged in literature. Obviously, this choice is arbitrary and offers both advantages and problems. The advantages of this strategy are that they not only allow one to overview an extensive literature in a limited space but it consents one to remain focused on the theoretical arguments, which are the main issue of this discussion. The problems are that this approach certainly implies the use of certain simplifications that cannot reflect the integrity of some arguments. To make explicit this strategy and its intentions may not solve the related problems, but it can draw attention to them and to the inherent limits of this approach. Then, if it is true that each of the theories that will be presented has a logical independence, and for this reason they will all be discussed separately, in many cases, they have been presented in various combinations.

The group of theoretical explanations that will be discussed in this section shares a common perspective: the idea that irregular migration is the result of states'

Table 3.1 The gaps

Discursive gap	**Irregular migration as states' choice**
Implementation gap	
Efficacy gap	**Irregular migration as states' failure**
Knowledge gap	

increasing inability to control international migrations. While this general idea is common to all of them, different positions have emerged regarding its extent. The most radical accounts have certified that states have lost control over their populations; in contrast, more nuanced ones have considered states to be still in control but in the process of weakening. There have been three main explanatory hypotheses as regards irregular migration being the result of states' ineffectiveness. A first approach has explained irregular migration as the result of the intrinsic and inevitable limitations of state mechanisms and policies. A second approach has focused on the role of those actors, forces and processes that, acting from outside the state, have been slowly eroding its prerogatives and control capacities. Finally, a third approach has concentrated on those actors, forces and processes that, acting from inside the state, have diminished its ability to manage migrant populations.

3.2.1 Intrinsic Limitations of States and Policies

Various scholars have explained irregular migration as the result of the internal, inescapable limitations that states experience concerning their control abilities. These interpretations have focused on the concrete difficulties found by states in developing effective mechanisms, systems and procedures to control a complex social phenomenon like migration. While the self-narrative built by modern states had envisaged the myth of absolute control over the population, in reality, even the most powerful and pervasive states have reached, at maximum, a high degree of control, but never total (Broeders, 2009; Van Meeteren, 2010).

As argued by Torpey: "in order to extract resources and implement policies, states must be in a position to locate and lay claim to people and goods" (Torpey, 1998, p. 244). In order to do that, states need not only to penetrate societies but also to "embrace" them. This latter metaphor that Torpey uses, highlights the complexity of the task; indeed, it is not only a question of setting up a bureaucracy or monopolizing the legitimate means of violence, but it is a matter of successfully registering all members of society and the main transactions that take place. As was discussed in Chap. 1, this effort by states to "enhance their grip on societies" (Broeders, 2009) has taken place in a very uneven way and has produced different results across history and geography. In this respect, Schrover and her colleagues have suggested that differences must be related to the particular processes of state formation in each case (Schrover, Van Der Leun, Lucassen, & Quispel, 2008). Other scholars have suggested that differences in the ability to control must be related to the different functioning and liberalness of the political system. However, in one of the first comparative analyses of irregular migration that includes non-western, non-liberal countries, Garcés-Mascareñas has concluded that also non-democratic administrations face important practical limitations to controlling their populations (Garcés-Mascareñas, 2012). As a matter of fact, if down through the twentieth century, states increasingly believed in their ability, "time showed that governments misunderstood the mechanisms that govern migration and overestimated the extent to which they were able to influence it" (Doomernik & Jandl, 2008, p. 20). Even after the migration crisis of the 1990s and the prioritization of migration control in the policy agendas, certain limitations have proved resilient. As a confirmation of this, the work of Broeders, for instance, after analysing the recent efforts made by Germany and the Netherlands, two among the most advanced and committed countries in the fight against irregular migration, concluded that those countries have not been "without setbacks and limitations" (Broeders, 2009, p. 193). Notwithstanding the huge investments, the implementation of the latest technologies and the diversification of policies (external and internal controls), in both cases it is still possible to identify what Broeders calls "white spots". This metaphor refers to those sectors of society and the economy that states, despite their efforts, cannot chart (Broeders, 2009, p. 194).

A number of specific reasons have been indicated to explain these limitations. Firstly, there are problems related to knowledge production and policy design; these imply a limited predictive ability, administrative loopholes, unintended consequences and policy failure (Bommes & Kolb, 2002; Freeman, 1995; Scott, 1998, 2008). Secondly, there are problems related to policy implementation, administrative competence and budgetary constraints (Broeders, 2009; Doomernik & Jandl, 2008; Massey, 1999; Scott, 1998; Van Der Leun, 2003). Just to give one example of this capacity problem, in the Netherlands, to reach the target of 10% of companies checked by labour inspectors to avoid irregular work, would require an increase in staff from the current 180 to 930 inspectors (Ministerie van Buitenlandse Zaken 2007 in Broeders, 2009).

From the perspective of these theories, irregular migration must be understood as an inevitable "fact of life" (Van Meeteren, 2010, p. 1), "a corollary of large-scale movements of people across national borders and governments' [imperfect] attempts to regulate migration" (Van Der Leun, 2003, p. 9). The merit of these approaches has been to relativize the myth of full control that characterized modern-state ideology and to show the limitation of state policies. They also called for a detailed and differentiated analysis of the administrative culture, methodologies and capacities of each state.

Although these are crucial aspects for the understanding of irregular migration, on their own, they have a limited explicative capacity. In particular, two issues remain beyond their grasp. Firstly, there exists the problem of policy intentionality: to acknowledge that control policies are imperfect and often fail, does not problematize the real aim of those policies or the possible conflict with other policies. It can be, for instance, that a certain degree of control failure and, thus, of irregular migration is the desired result or an acceptable compromise among multiple objectives. Secondly, the approach overstates the capacity of policies and does not consider other factors. For instance, policy limitations can help to explain why, under heavy migratory pressures, controls may fail, but they do not say much about the reasons for, and the variability of, those pressures; the same state with identical control policies in a certain historical moment may experience high levels of irregular migration and in others very low levels.

3.2.2 *External Constraints of States and Irregular Migration*

The theoretical arguments that will be discussed in this section concentrate on a number of factors and processes, mainly external to states, which have contributed to the erosion of their ability to control populations. These have been interpreted as the main cause of irregular migration.

The Effects of Globalization: Economy, Politics and Society

An extensive literature has linked the increased relevance of irregular migration with the effects of globalization. In particular, many interpretations have found in the complex and multileveled transformations brought by globalization the reason for states' increasing weakness and ineffectiveness in controlling international migration. The argument here has been that the particular characteristics of the current age are undermining states' capacities and that irregular migration is only but one of the signs of this process. The use of the concept of globalization and the problem of a definition could open a way to a very interesting, but probably endless, debate. As for this discussion, a very broad and general definition will be used: "globalization can be defined as the intensification of worldwide social relations which link distant localities in such a way that local happenings are shaped by events occurring many miles away and vice-versa" (Giddens, 1990, p. 64). While in literature it is possible to find many different approaches that correlate irregular migration and globalization, three main general arguments seem distinguishable.

Economic Globalization and Irregular Migration

Many scholars have linked the current trend of irregular migrations to the far-reaching economic transformations that have affected both migration-sending societies and receiving ones in the last decades. These transformations would appear to have determined a sharp increase in the number of potential migrants in the former case and a substantial rise of the demand for migrant labour in the latter one. The combination of these two circumstances, in other words the simultaneous intensification of push and pull factors, would seem to have determined a powerful support for international migration. Overwhelmed by the dynamics of these economic forces, states' attempts to limit migrations have proved ineffective: when they tried to close regular entry channels, migrants shifted to the irregular ones.

Research on these processes has followed two key lines. A first group of scholars has focused on the systemic, international transformation of the global economy. Different approaches and theories, with various degrees of politicization, have emerged on this. The general argument has been that the rapid and worldwide diffusion of the free market principles has determined a sharp transformation in the functioning of the economy. Whereas, up until the 1970s, states, using monetary, commercial and other regulatory policies had been able to control and govern the main economic transactions, since then, an increasingly more integrated and autonomous global market has been developing. Various concepts and historical labels have been used to describe this process, for instance, deregulation, flexibilization, Washington consensus, neoliberal globalization, etc. The impact of these wide-ranging transformations has been twofold: on the one hand, the rapid dismantling of traditional economies in non-industrialized countries; on the other, the restructuring of the Fordist economy and social model in industrialized ones. The joint effect has been an enormous increase of global interdependence and a continuous rise in the exchange of goods, capital and information. The leading force is now the law of demand and supply and the means of production has had to follow profit opportunities rather than states' desires or plans. This process of deregulation and increasing economic interconnectedness has had an inevitable corollary, a strong reinforcement of population movements (Broeders & Engbersen, 2007; Castles, 2010a; Cornelius et al., 1994; Massey, 1999; Massey et al., 1998; Sassen, 1998; Schierup, Hansen & Castles, 2006). From a purely economic point of view, in fact, labour is a means of production, just like capital, raw materials or machinery; if globalization implies the free movement of capital, raw materials and machinery in search of the most profitable production conditions, the same must work also for labour. The conditions for irregular migration to emerge as a structural phenomenon of globalization would be located in the paradoxical fact that, while states have largely accepted economic interconnection and the free flow of other means of production, they have fiercely opposed the free flow of workers (Cornelius et al., 1994; Cornelius & Tsuda, 2004; Guiraudon & Joppke, 2001; Hollifield, 1992, 2004). In a context of growing interdependence, powerful mobility forces, but limited regulatory capacities, states find it increasingly difficult to maintain their control stance and to avoid irregular entries.

A second group of scholars has focused more on the effects of globalization in the receiving-country economies. In particular, they have stressed that the process of economic restructuring that followed the economic crisis of the 1970s has radically transformed both the production structures and the labour conditions in industrialized countries. While up until then, a largely protected economy was the basis for a unified labour market, widespread labour rights and stable, unionized employment relations, the erosion of the Fordist model and the opening up of the national economy to international fluxes had a disrupting effect. A number of processes have been analysed: the development of dual-labour markets (Piore, 1980; Portes, 1978; Sassen, 1998); the flexibilization, deregulation and informalization of many sectors of the economy (Castles & Miller, 1993; Sassen, 1998; Schierup et al., 2006); the decline of many industries and the process of delocalization (Portes, 1978; Sassen, 1998; Schierup et al., 2006); the drop in unionized labour (Castles, 2004); the development of subcontracting (Baldwin-Edwards & Kraler, 2009; Broeders & Engbersen, 2007; Martin & Miller, 2000); the rise of specific urban informal economies (Sassen, 1998; Van Der Leun, 2003). On the whole, these processes have determined an increasing demand for a cheap, unqualified and flexible work force. Since native workers have generally not been willing to accept the new working conditions, this demand turned to migrant workers. As discussed above, the combination of a high demand for migrants and a political reluctance to accept them, made irregular migration a somehow "natural" solution to the mismatch. Moreover, as many scholars have underlined, irregular migrants, because of their precarious conditions, have fulfilled in an optimal way the demand of many sectors of the economy. They are a cheap, hyper-flexible, unprotected and extremely mobile resource (Castles, 2010a; Sassen, 1998). As stated by Hanson: "illegal immigration is a persistent phenomenon [...] because it has a strong economic rationale" (Hanson, 2007, p. 32).

Political/Legal Globalization and Irregular Migration

Another strand of research has concentrated on the political effects of globalization that have affected states' capacity to control and deter international migration. Here the lines of research have followed two main paths. On the one hand, from a more theoretical point of view, it has been stressed how "globalization transcends the territorial borders of states and, as a consequence, profoundly affects the nature and functions of state of governance in the world political economy, including of course, the governance of migration" (İçduygu, 2007, p. 145). Many different processes have been studied: the increasing international anomie (İçduygu, 2007); the fluidity and openness of contemporary societies (Urry, 2007); the process of de-territorialisation that implies a weakening of state borders and sovereignty (Friese & Mezzadra, 2010); the interconnectedness and interdependence of the world system (Wallerstein, 2004). From this perspective, irregular migration is one of the phenomena that indicates more clearly how globalization is determining the erosion of states' prerogatives and, in particular, their sovereignty. What is at stake is not only the economic functioning of the international order, but the political one. While states and borders had been the cornerstone of the Westphalian system, the uncontrolled global fluxes of the contemporary era are the concrete evidence of its decline.

On the other hand, many scholars have discussed how the development of an international human-rights regime (Cornelius et al., 1994; Guiraudon & Lahav, 2000; Jacobson, 1996; Sassen, 1998; Soysal, 1994) and of an international framework of institutions (Cornelius & Tsuda, 2004; Geddes, 2001) have strongly limited, from the outside, the ability of states to control and govern their populations. From this perspective, the obligation for states to compel to a number of international agreements and treaties that protect the rights of migrants both as they move across borders and once they arrive inside hosting societies, has greatly constrained states' restrictive power. Moreover, the increasing importance and influence of international institutions and the development of agencies specifically focused on migration, like IOM and UNHCR, have also concurred on the limitation of states' arbitrariness and on the creation of a shared system of safeguards for migrants. Within this context, the ability of the latter to bypass, circumvent and evade state controls has grown enormously. For instance, the widespread guarantee of asylum rights, the *non-refoulement* principle, the right to appeal asylum rejection and expulsion orders, the possibility for origin countries to refuse the re-admission of non-identified migrants, not only have empowered migrants vis-à-vis states but, through their misuse, have offered a number of opportunities for irregular migration.

Social Globalization and Irregular Migration

Finally, a number of scholars have focused on the social implications of globalization and their impact on migration trends and on state control capacities. These analyses have highlighted how globalization has concurred to strengthen the social dynamics of migration. As pointed out by Castles, globalization has offered the technological and cultural basis for mobility to increase and involve all regions of the planet (Castles, 2010a). Although networks, cumulative causation, social capital, and chain theories have always had an important role in explaining migrations (Castles & Miller, 1993; Massey et al., 1998), in the context of globalization, these approaches gained particular relevance. The improvements in communication and transport systems, from this perspective, have greatly reinforced the self-perpetuating characteristics of migration and, therefore, one of the crucial elements particularly of irregular migration (Castles, 2004; López Sala, 2005; Massey, Goldring, & Durand, 1994). As discussed by López Sala, the impossibility for irregular migrants to count on formal channels and the increased difficulty of their migratory process make their reliance on networks and social capital an indispensable asset for their success (López Sala, 2005). From this viewpoint, the transformations brought about by globalization have offered migrants new and more sophisticated tools that enable them to share information, develop strategies and effectively contrast state controls.

A similar argument, but with a stronger theoretical ambition, has been developed, especially since the mid-1990s, through studies on migrant transnationalism (Faist, 2000; Glick-Schiller, Basch, & Blanc-Szanton, 1992; Glick-Schiller, Basch, & Szanton-Blanc, 1995; Portes, 2001, 2003; Portes, Guarnizo, & Landolt, 1999). The idea is that the development of migrant networks in the context of globalization is not merely easing migration processes, but is actually leading to the development of real transnational communities. These, in their turn, are increasingly capable of transcending state borders and challenging principles, such as, membership, citizenship, and sovereignty (Castles, 2004). Within this context, irregular migration appears as a correlated phenomenon that clearly exemplifies the contradiction between the old statist organization of space and populations, and the new, emergent, transnational one.

The main virtue of all these theoretical explanations has been to pinpoint those broad and far-reaching transformations brought about by globalization that are affecting states' capacity to control international migration. From this viewpoint, irregular migration is essentially the result of a structural conflict between global forces pushing for an ever-greater interconnection and flux of information, goods, capital and people, and states. While these theories offer a framework to understand the current general trends of irregular migration, when it comes to the interpretation of specific irregular fluxes and populations they are less useful. How can they explain the highly differentiated picture? This difficulty is probably related to the fact that they have too easily dismissed the role of states. As a matter of fact, while it is true that irregular migration has become a widespread phenomenon, important disparities exist between different national contexts. In this sense, the question to be answered would be: how are globalization processes interacting with the different social, political and economic contexts and how do different forms of irregular migration emerge from these particular interactions?

The Irregular Migration Industry

A number of scholars have linked the increased inefficiency of state policies and the increasing prominence of irregular migration to the emergence of the so-called "migration industry" (Andersson, 2014, 2016; Castles, 2004, 2010a; Castles & Miller, 1993; Koser, 2010; Kyle & Koslowski, 2011; Zolberg, 2006). As put by Castles and Miller, the term: "embraces the many people who earn their livelihood by organizing migratory movements" (Castles & Miller, 1993, p. 114). These "people" include a wide variety of actors that range from migrant community members, to small informal entrepreneurs, to actual criminals often connected to international mafias (Kyle & Koslowski, 2011). They support, back up and often exploit migrants along their journey in exchange for money. The services provided include for instance: lawyers who advise on how to circumvent laws and controls, human smugglers that help migrants to cross the borders, false document providers, labour and housing providers, credit providers and usurers, etc.

From this perspective, the services offered by this background support network have become crucial to circumvent controls and thus to make irregular migration possible. This has become especially true since control efforts by states dramatically increased in the aftermath of the so-called migration crisis of the 1990s. Whereas in the previous phase, many relatively easy entry channels existed for irregular migrants and the use of personal networks was often enough for success, the efforts made by states to enforce borders and close the main legal loopholes changed the scenario. In the new context, spontaneous irregular migration turned increasingly ineffective and the recurrence to "professional" services became indispensable. This, in turn, created a whole new range of entrepreneurial opportunities and raised the related profit margins, generating the development of a truly global "migration industry" (Andersson, 2014). Today, the enormous economic interests involved and the extension and relevance of this industry can hardly be underestimated and it certainly provides a powerful explanation for the difficulties experienced by states in controlling migratory movements. As expressed by Harris, this has become "a vast unseen international network underpinning a global labour market; a horde of termites… boring the national fortification against migration, and changing whole societies" (in Castles & Miller, 1993, p. 115).

The uncovering of the importance of the migration industry has provided another important explanation of irregular migration. The difficulties experienced by states in effectively controlling their borders and curtailing irregular fluxes has depended not only upon the individual efforts made by migrants, or upon their turning to networks and personal contacts, but also, and increasingly so, upon a powerful industry that has supported and encouraged migrants' efforts. While this claim is unquestionably important and has been supported by relevant evidence, two critical aspects may be raised. To begin with, caution must be used regarding the arrival at conclusions derived from it. While it is true that states have had difficulties in controlling irregular migration and this can be related to the migration industry, the fact that the phenomenon is, nevertheless, limited shows that states have not lost control. Furthermore, while the role of the migration industry is an important piece of the irregular migration puzzle, on its own, it does not provide much explanation. The dissimilar social and numerical relevance of the irregular migration phenomenon in different countries shows that the effects of the migration industry are not the same. Why is this happening? Why, for instance, are certain states more effective than others against human smugglers? Or why does the same country experience different levels of irregular entries at different times? These critiques seem to point to the fact that the migration industry plays a crucial role as a catalyst for irregular migration fluxes once they have started.

3.2.3 *Internal Constraints of States and Irregular Migration*

In this section, the theoretical arguments that have concentrated on those factors and processes, acting mainly inside state territories, that have contributed to incrementing the demand for migrants, to the erosion of state capacities to control population movements and, hence, to the development of irregular migration will be discussed.

The Role of the Informal Economy

Though a number of links between the current economic trends and irregular migration have already been addressed, the relevance given in literature to the role of the informal economy demands for a separate discussion. In this section, the focus will be placed on those approaches that understand the informal economy as a sign of current erosion of state prerogatives. Irregular migration, from this perspective, would then be a consequence of those forces that, from the inside, limit the regulatory capacities of states.

Throughout the 1980s and 1990s, the idea that the world's economies were on an ineluctable path to "modernization" and, thus, to "formalization" appeared increasingly questionable. Even in the most advanced countries, where for some decades the "formalization thesis" (Williams & Windebank, 1998) seemed to apply, signs of an opposite movement were increasing. "What is new in the current context is that the informal sector grows, even in highly institutionalized economies, at the expense of already formalized relationships" (Castell and Portes 1989, p.13 in Samers, 2004, p. 2003). This development was linked to various factors: the necessity for employers to reduce costs and increase flexibility; the "care deficit" created by native female employment; the transformation of urban economies and the emergence of ethnically-specialized sectors (Samers, 2004). More in general, as argued by Sassen, informalization must be seen in the context of the economic restructuring that has contributed to the decline of the manufacturing-dominated industrial complex of the post-war era and the rise of a new, service-dominated economic complex (Sassen, 1998). Many scholars have pointed to this process of informalization as an explanation of the rising significance of irregular migration in receiving countries. From this standpoint, the informal economy works as a magnet for irregular migrants, as it offers them the possibility to find employment (Baldwin-Edwards, 2008; Düvell, 2006; Quassoli, 1999; Reyneri, 1998, 2004). As pointed out by Sassen (1998), immigrants "may be in a favourable position to seize the opportunities presented by informalization, [...] but they do not necessarily create such opportunities. Instead, the opportunities may well be a structured outcome of the composition of advanced economies" (Sassen, 1998, p. 154).

The theories that have focused on the role of the informal economy have offered a convincing argument to explain the demand for irregular migration. The main advantage of this approach has been that it directly links the phenomenon to the particular social and economic configuration of each national context. In this sense, it calls for a differentiated analysis of the structural conditions that may favour irregular migration or not. This perspective, nevertheless, has not been free from flaws. On the one hand, the relation between irregular migration and the informal economy cannot be linearly interpreted and does not necessarily indicate state failure. In many countries the informal economy had been an internal characteristic long before the arrival of migrants. Moreover, a number of national studies have shown that states do not always put all their efforts into controlling the informal economy but display, instead, tolerant attitudes (Jordan et al., 2003; Reyneri, 1998; Triandafyllidou, 2009). In this sense, the informal economy alone cannot explain irregular migration and it does not necessarily imply the erosion of state prerogatives. On the other hand, studies in many countries have shown that irregular migrants do not necessarily rely on the informal economy. A notable case concerns the US that has one of the smallest informal economies in the world (Schneider, Buehn, & Montenegro, 2010), but a sizable number of irregular migrants (Passel, Cohn, & Gonzalez-Barrera, 2013). This is possible because there is a limited enforcement of labour controls and, therefore, a tacit tolerance of the regular work of irregular migrants. These examples show that, given the great variety around the world of economic arrangements, ways and degrees of law enforcement, and levels of toleration of informality, the explanation of irregular migration requires differentiated and customized analysis.

The Role of Migrants' Agency

Departing from a critique of the structuralist explanations of irregular migration, an important line of research has focused on the role of migrants' agency. From this perspective, the excessive emphasis laid on state policies or on the economic dynamics has neglected the crucial importance of migrants' actions and strategies. Migrants are not passive recipients of policy measures or victims of capitalist logics; on the contrary, they are active players who are perfectly capable of analysing the opportunity structure they encounter, of developing strategies and of circumventing restrictions. From this viewpoint, irregular migration has been explained as the result of these capacities and of the ability of migrants to exploit the loopholes and weaknesses that characterize state controls.

The theoretical explanations that have centred their attention on the role of migrants have provided different accounts on the extent to which they are able to confront and challenge social structures. For some scholars, migrants' agency has mainly a reactive and, on the whole, a limited capacity to defy structural forces. The attention has focused on the concrete strategies that enable an irregular migrant to "survive" within a very limited range of possibilities. For other scholars, migrants' agency is a much more powerful force that is able to transgress, contest and even modify social structures. Here, importance has been given to the strategies developed by irregular migrants, to their political activism, and to the social and political transformations they are backing.

Focusing on the first group, there have been many approaches and findings. Espenshade has suggested that irregular migrants see policy barriers as one of the obstacles of the equation they face once they decide to migrate (Espenshade, 1995). In this sense, they have a very pragmatic approach: they estimate difficulties, consider alternative options, share information and take decisions. To do so, they extensively count on the use of formal and informal networks, which in their case play a fundamental role (Broeders & Engbersen, 2007; Engbersen, 2001; Engbersen & Broeders, 2009; Portes, 1978, 1996). Paradoxically, it may happen that irregularity is an advantage over regular migration (Bommes & Sciortino, 2011; Schrover et al., 2008). Indeed, in certain contexts, being irregular offers better economic opportunities or more flexibility and the possibility to elude state controls (Garcés-Mascareñas, 2012). As regards the specific strategies developed by irregular migrants, a diversified picture has been sketched. Engbersen has identified six strategies: the mobilization of social capital, bogus marriages, manipulation of identity, strategic operations in the public space, legal action, and crime (Engbersen, 2001). As for the manipulation of identity, there are three main tactics: false identity adoption, destruction of documents, and concealment of irregular status (Engbersen, 2001). Vasta has concentrated on the functioning of the paper market. She has shown how irregular migrants engage in a dialectic process with the structures and control mechanisms of receiving societies. Buying, renting, or borrowing someone else's papers is part of a productive process by which migrants permanently construct and re-construct their subjectivity (Vasta, 2011). Van der Leun, working on the Dutch case, has shown how irregular migrants are able to find and actively exploit the loopholes that characterize the legislation and the implementation of control policies (Van Der Leun, 2003). On the one hand, the complexity of legislation, the different dimensions and sectors of application, and the existence of various and often uncoordinated levels of governance determine the presence of legal ambiguities, contradictions and voids. On the other, irregular migrants and their networks, often with the help of lawyers, NGOs and even street-level bureaucrats, successfully learn to take advantage of these pitfalls. Another type of strategy is to resort to sectorial shifts or even to criminality to avoid labour enforcement (Engbersen & Van Der Leun, 2001). De Haas (2011) has identified four main substitution effects that limit the effectiveness of restrictions: spatial substitution (moving to other regions or other countries in search of better opportunities); categorical substitution (reorientation towards other legal or illegal sectors to avoid controls); inter-temporal substitution (modifying the timing and length of migration); reverse flow substitution (the adoption of return migration when restrictions decrease) (de Haas, 2011).

Regarding the second group of studies, a number of concepts have been proposed to capture the broader social significance of irregular migrants' networks and strategies. The intention of these approaches has been to underline the social and processual character of irregular migration (Castles, 2010a). Hughes, for instance, has proposed the notion of "bastard institutions" (Hughes 1951/1994 in Leerkes, 2009), and Mahler that of a "parallel institution" (Mahler, 1995). More recently, Bommes and his colleagues have used the concept of "foggy social structures" to indicate those "social structures that emerge from efforts by individuals and organizations to avoid the production of knowledge about their activities by making them either unobservable or indeterminable" (Bommes & Kolb, 2002, p. 5; Bommes & Sciortino, 2011; Engbersen & Broeders, 2009, p. 868).

Following this orientation, in a recent work, Van Meeteren has enquired into how the different aspirations of individual irregular migrants determine differentiated patterns of insertion in the host societies (Van Meeteren, 2010). From her perspective, the concrete experience of irregular migrants cannot be understood only on the basis of the structural conditions they encounter. Indeed, she states: "contexts do not mechanically constrain or construct irregular migrants' actions. Instead, they take advantage and react to the window of opportunity in different ways" (Van Meeteren, 2010, p. 31). To fully grasp their experience, it is necessary to include in the analysis migrants' agency and, in particular, the role of aspirations. Researching on the case of irregular migrants of different nationalities in Belgium and the Netherlands, she identified three main types of aspirations: settlement (the goal is to settle in the host society), investment (the goal is to save money in order to return to the origin country), and legalization (the goal is to regularize the status in order to start a new life). The different aspirations not only translate into diverse strategies and ways of interaction with the host society on the part of migrants, but also into very different outcomes in terms of living standards, degrees of incorporation and social relations. The study shows how, within the same structural context, the three different types of aspiration transform into clearly distinguishable forms of incorporation both in its functional (housing, work, sources of income) and its social dimension (leisure time and social contacts). "Investment migrants" concentrate on working hard, saving money and planning the return home. Accordingly, they: work as much as they can, accepting bad jobs rejected by natives since they see them as temporary; are usually alone and spend as little money as possible, living in bad conditions and in degraded districts; do not value leisure time and when not on duty, stay at home; have very small networks of social contacts and maintain many connections with the origin country. "Settlement migrants", instead, assume that the receiving society is their new home. Hence, they: prefer stable, non-seasonal work with free time often in native households; have families with them and are willing to spend more on better housing in residential suburbs; travel around, spend on leisure and maintain an intense social life; have large social networks in the host country and limited contacts back home. Finally, "legalization migrants", whose main objective is to regularize their status, lead a very particular life. They: work as little as possible due to the risks of compromising their aspirations; rely on natives and organizations rather than on their communities to get support, since they do not work; have a lot of free time that they spend elaborating their strategies to legalize their situation (marriage strategy, legalization strategy); have limited contacts with their origin countries and do not remit. This analysis leads Van Meeteren to conclude that "overemphasizing structure in the analysis obscures understanding of how migrants act differently under similar circumstances because they have different aspirations" (Van Meeteren, 2010, p. 135).

Another interesting standpoint within this line of enquiry has been advanced by a number of scholars who, in recent years, have developed the "autonomy of migration" perspective (Mezzadra, 2011; Papadopoulos, Stephenson, & Tsianos, 2008; Papadopoulos & Tsianos, 2007). Their approach does not consider migration in isolation from social, cultural and economic structures; in fact, they consider that "the opposite is true: migration is understood as a creative force within these structures" (Papadopoulos et al., 2008, p. 202). The main objective of these scholars, as pointed out by Mezzadra, has been that of: "..looking at migratory movements and conflicts in terms that prioritize the subjective practices, the desires, the expectations, and the behaviours of migrants themselves. […] It allows for an analysis of the production of irregularity not as a unilateral process of exclusion and domination managed by state and law, but as a tense and conflict-driven process, in which subjective movements and struggles of migration are an active and fundamental factor. […] The autonomy of migration looks at the fact that some migrants, both regular and irregular, act as citizens and insist that they are already citizens (Mezzadra, 2011, p. 121). For these authors, the agency of irregular migrants does not simply allow them to solve their basic problems or to cross borders. Instead it should be read as a force that is able to challenge the legal frameworks and institutions built by states and, in so doing, concurs with their transformation. In this regard, particular attention has been given to the relationship between irregularity and citizenship. Whereas the latter has been usually interpreted as a unilateral concession by the state and, thus, as a tool of domination and control from above, the autonomy of migration perspective, recalling the work of scholars, like Balibar, Isin or Honig, has proposed a more dynamic and dialectic understanding of it. Citizenship must be considered as an 'institution in flux' (Isin, 2009), as a political/legal arrangement that is permanently contested and modified by the interplay of old and newly-emergent social forces (Balibar, 2001; Honig, 2009; Isin, 2009; Mezzadra, 2011).

The main contribution of the theories presented in this section has been the shift of focus away from the structural contexts to illuminate the crucial role of migrants' agency. In particular, the theories have warned against the tendency to uncritically accept the narratives that postulate the state as the main and undisputed actor within society, and migrants as passive victims of its dispositions. Irregular migration, from this perspective, is precisely one of those phenomena that reveal the limits of politics in deciding and controlling social life. The different accounts have shown how the individual and cooperative actions of migrants have been able to challenge state decisions, barriers and goals. The extent to which this has been made possible was interpreted in different ways, ranging from those authors who acknowledged a limited, mainly adaptive capacity to those who described a significant and potentially transformative one. Though the contribution of these approaches has been crucial to obtaining a more comprehensive understanding of irregular migration, a number of critical points can be identified. On the one hand, there has been a problem with the emphasis given to the agency argument. The necessity to amend the excessive attention given in literature to structural explanations has often turned into excess in the opposite direction. The focus on migrants' strategies, networks and aspirations in many cases has led to downplaying the role of structures, especially of politics and economics. In particular, the accounts that have ascribed a wide-ranging transformative ability to irregular migration and have described the state as a zombie category appeared to be unrealistic (Schinkel, 2009). If it is indeed true that irregular migration reveals the limits of controls and the relevance of individuals' choices and actions, this cannot be linearly interpreted as the failure or the irrelevance of politics. Both the confined character of the phenomenon and the generally harsh conditions that irregular migrants experience indicate that the role of the state is far from marginal. Moreover, as will be shown, the hypothesis of states' fierce antagonism towards irregular migration cannot be uncritically accepted, since state ambiguities have been widely documented. On the other hand, the tendency to detach the analysis from the structural contexts has frequently generated broad conceptualizations of irregular migration as a general and undifferentiated phenomenon. The empirical research, however, has consistently shown that irregular migration assumes different shapes and characteristics within the different contexts. Moreover, even within a single context, a change in the structural conditions has been shown to determine changes in the strategies enacted by migrants or even in their aspirations (Van Meeteren, 2010). These examples show that only a dynamic and interactive understanding of the relationship between structures and agency can offer an adequate framework to conceptualize irregular migration.

Internal Social Constraints

Finally, another important line of reasoning has emphasized how a variety of actions, decisions and initiatives taken by actors internal to the hosting society have concurred to the ineffectiveness of control policies and, therefore, to the irregular migration phenomenon.

A number of scholars have focused on the ways in which the policies are actually implemented at the lowest levels of the administration (A. Ellermann, 2010; Jordan et al., 2003; Lahav & Guiraudon, 2006; Scott, 1998; Van Der Leun, 2003). Their researches have enquired into the activity of police officers, public service employees, social workers, healthcare and education workers, etc. Their analyses have generally revealed the existence of important margins of discretion in the application of written laws and of "a pluralistic and multi-layered system of actors who have their own deliberations and professional considerations" (Van Der Leun, 2003, p. 173). Many reasons have been proposed to explain this phenomenon. Van der Leun, writing on the behaviour of street-level bureaucrats vis-à-vis irregular migrants in the Dutch case, has evidenced five: (A) the professional morale and degree of discretion (for instance, doctors may give priority to saving a life rather than to the application of a restrictive law); (B) the degree of face-to-face contacts with clients (a more personal contact generally leads to higher degrees of lenience); (C) the availability of alternatives on the market (in the sector of social housing and adult education for instance, irregular migrants can easily be referred to private landlords or to community centres); (D) the costs (the higher the costs of the services provided will probably mean more restrictive decisions); (E) the interference with other policies and duties (for instance, a police officer may have to prioritize arresting criminals rather than irregular migrants) (Van Der Leun, 2003). Jordan, Stråth and Triandafyllidou have shown how different organizational cultures may determine a different mediation between the top-down pressures from politics and the bottom-up pressures from migrants themselves, local employers and communities, and from non-government organizations (Jordan et al., 2003). In a similar vein, Cornelius and Tsuda have stressed the importance of the national political culture in determining different efficiency standards in policy implementation (Cornelius & Tsuda, 2004). All these contribution have highlighted the importance of the local social context in determining the conditions and opportunities for irregular migration to exist. As stated by Van der Leun: "the very reason that illegal immigrants can circumvent or bypass legal limits, is that loopholes come into existence when local actors have, at least partly, different considerations than proponents of full exclusion or restriction" (Van Der Leun, 2003, p. 174).

Enquiring into the internal limitation to migration controls, another strand of research has focused on the different types of support that irregular migrants find within the host society. Considering what has been referred to as "the ecology of illegal residence" (Leerkes, 2009), two main types of support seem to be clearly distinguishable. On the one hand, there are the services provided on a lucrative basis by what can be considered the internal counterpart of the migration industry. On the other hand, there are the services provided on a free basis by civil society institutions, NGOs, charity organizations, etc. In the first group, research has focused not only on the role of informal employers, as was discussed regarding the role of the informal economy, but also on a whole galaxy of actors that offer their services to irregular migrants in exchange for money. These include: fake document suppliers, housing providers, doctors, teachers, lawyers, bogus marriage arrangers, etc. Within this group, criminal networks may also play a part. As shown by Engbersen and his colleagues, when the other channels and opportunities are closed, irregular migrants may be forced to turn to criminality in order to find the means to survive (Broeders & Engbersen, 2007; Engbersen, 2001; Engbersen & Broeders, 2009; Engbersen & Van Der Leun, 2001; Leerkes, Van Der Leun, & Engbersen, 2012). Within the other group, research has likewise evidenced the existence of a great variety of actors and institution within the so-called civil society that help irregular migrants in many ways. Within these, some have been more concerned with offering material support like shelter, food, legal assistance, etc.; others have adopted a more political stance, focusing on helping irregular migrants to organize protests, present instances, claim their rights, etc. In this respect, however, it has been underlined that the attitudes towards irregular migration, and therefore the support, may sharply differ from one social context to the other. Not only may this be the case, but, as stressed by Düvell, within each society it is possible to find members that support, tolerate or ignore the irregulars. In this sense, one should consider that often "the moral of the community differs from the law" (Düvell, 2011, p. 63). Both these types of support have concurred, though in different ways, to make the residence of irregular migrants possible in their hosting societies, especially where highly restrictive and excluding policies have been enacted.

The discussion on the internal constraints to migration control has underlined how a number of factors determine a state's impossibility to fully and thoroughly control all social transactions. This has been the result of both the difficulties and inconsistencies of policy implementation, and the independence and unconformity of many social actors from the legal and moral stances of states. For irregular migrants, this has transformed into a number of opportunities and sources to rely on, for making a living even within very restrictive contexts. The main contribution of these approaches to the theoretical understanding of irregular migration has been twofold. It has evidenced the complex functioning of the political processes and the fragmented, multi-levelled character of the state. From this perspective, an adequate understanding of irregular migration requires going beyond a legalistic approach and demands for an analysis of the actual implementation of the laws. It has also emphasized the social character of the phenomenon, which implies that policies do not act within an empty space, but within a complex web of actors, institutions and interactions that display contrasting interests. The way in which the irregular migration phenomenon configures is not the straightforward result of policies, but, instead, of the interaction of them with the rest of society. The critical aspects of these arguments concern the extent to which they are used to sustain the idea of states' diminished capacity to control migrations. Although both the main arguments presented certainly raise attention to the difficulties experienced by states in making their goals effective, this does not mean they are powerless or have lost control over their populations.

3.3 Irregular Migration as Choice of States

The second group of theories that will be discussed departs from a reverse evaluation of policy efficacy and state capacity to control international migration. Policies are effective and states are really capable of governing migration fluxes and populations (Brochmann & Hammar, 1999; Caplan & Torpey, 2001; Freeman, 1995;

Guiraudon & Lahav, 2000; Joppke, 1998c). While this position was central, especially in Marxist interpretations of irregular migration since the 1970s, it re-emerged with new strength in the second half of the 1990s to contrast the chorus of voices that had sentenced the state to death too early. Rather than losing control or being a zombie category, states have been perceived as successfully adapting to internal and external pressures through the development of new strategies and increasingly-effective control mechanisms. If irregular migration exists, this does not indicate a failure on the part of the state but, rather, an explicit or implicit choice in this direction. In this interpretation, the whole conceptualization of irregular migration radically shifts: the question is no longer why migration control fails but, why states decide to allow or not to allow certain levels of irregular migration.

A variety of theoretical explanations and have emerged. Among these, it seems possible to identify two very different perspectives. The first has understood irregular migration as a by-product of the particular configuration and functioning of modern states. The focus has been placed on the analysis of the different functional imperatives of the state and on the ways these are fulfilled. The second group, instead, considering the state mainly as a broker, has concentrated on the different interests connected to irregular migration present in society and on the ways they are articulated to become relevant for politics. Irregular migration, from this point of view, is "produced" or "allowed" by the state, depending on the viewpoint, in order to respond to the ever-changing equilibrium among the different social demands.

3.3.1 State Imperatives and Irregular Migration

The interest in the internal structures and functioning of states and in the way these have an influence on irregular migration has followed a number of different paths. In particular, there have been three main arguments proposed in literature. The first has centred on the analysis of the concept and functioning of sovereignty and has found in this fundamental institution of modern states the main explanation of irregular migration. The second has directed its attention towards the relationship between the state and populations and the different techniques developed by the former in order to control the latter. The third has focused on the particular institutional configuration and functioning of liberal-democracies and has explained irregular migration in relation to the self-restraint of the state as regards control policies.

State Sovereignty and Irregular Migration

The relation between sovereignty and control of populations has always been a central issue both in migration and political-theory literature. However, the topic received renewed interest in connection with the debates around globalization and the migration crisis of the 1990s.

The works of Agamben have offered a particular interpretation of this relation that proved to be particularly influential in the subsequent decade (Biswas & Nair, 2009; Broeders, 2009; De Genova, 2002, 2010; A. Ellermann, 2010; Antje Ellermann, 2014; Schinkel, 2009, 2010). His starting point was precisely that of contesting the widespread idea that modern states had a naturally-granted and unproblematic authority to control their territories and populations. Rather than an intrinsic and inalienable property or a transcendentally-derived authority, Agamben sees sovereignty as a power that must always be reaffirmed and which is, then, always at risk (Agamben, 1998, 2003). Reflecting on the work of Carl Schmitt, who defined the sovereign as the actor "who decides on the exception" (Schmitt, 2008), Agamben identifies in the "state of exception" the fulcrum on which the whole structure of sovereignty, and thus of state power, is built. Accordingly, if sovereign power is the ability to establish what is exceptional to an order, sovereignty is the logic by which such an order comes into being (Biswas & Nair, 2009, p. 5). However, instead of understanding these concepts in an abstract, juridical perspective, Agamben argues that it is possible to observe the logic of sovereignty at work in multiple sites at any time. In his works, he has scrutinized history in search of paradigmatic cases of "states of exception". In his view, the Nazi camps or the Guantanamo prison are perfect examples of the sovereign power deciding to suspend the common order in order to reaffirm its power (Baldwin-Edwards, 2008; Balibar, 2010; Caplan & Torpey, 2001; Cornelius et al., 1994; Torpey, 1998).

In *Homo Sacer. Sovereign power and bare life*, his attention focuses on the distinction made in ancient Greek between the concepts of *zoe* and *bios*. Though both terms generally mean *life*, the former refers to it as the basic, biologic, "bare" existence shared by all living creatures, while the latter refers to the politically-qualified, characteristic existence of a specific people within a certain order. For Agamben, the production of *bios* and its distinction from *zoe* is the "original activity of sovereignty". Only the banning and the exclusion of *zoe* from the political community enables the distinction from *bios* and, therefore, justifies the existence of the sovereign. Yet, since the sovereign power is constituted by the exclusion of *zoe*, the complete alienation of this would eliminate the reason for being of such power. That is why *zoe*, in order for the sovereign power to hold its significance, must be included in the sovereign realm as excluded "bare life". In this sense, "the banishment of bare life by sovereign power, which excludes it from all political life and denies it any juridical validity", still implicates "a continuous relationship" (Agamben, 1998; De Genova, 2010, p. 37). The irregular migrant is the figure that best incarnates the concept of *zoe* as opposed to the one of *bios*, the citizen. He or she represents the "bare life" whose exclusion enables the existence of the citizen, and so legitimizes the role of the state. In the words of Agamben: "It is the exclusion of bare life on which the *polis* rests" (Agamben, 1998: 13). The detention centres for migrants are proof that "the camp" is not a historical anomaly, but the "hidden matrix of our time", "the *nomos* of the political space in which we still live" (Agamben, 1998). If every migrant would in-*mediately* (hence, without mediation) become a citizen and hold the same rights as a citizen, the very power that "mediates" and gives meaning to the distinction would become powerless and, therefore, meaningless. The irregular migrant is then the fundamental antagonist of sovereignty but, at the same time and for the same reason, its most necessary counterpart.

In a similar fashion, Schinkel has pointed out that "the so-called 'problem of illegality' is but one expression of a problem of self-maintenance of the society/nation-state dichotomy in times of globalization and system integration" (Schinkel, 2009, p. 790). In his view, the state, defied by the forces of globalization, and, in particular, by the declining relevance of space, is trying to redefine itself in order to survive. Recalling Agamben's concept of *the camp*, Auge's concept of *non-places* and Foucault's concept of *heterotopia*, he emphasizes precisely the spatial component of this redefinition. Through the incarceration of irregular migrants in detention centres and their eventual expulsion, the state is able to reintroduce and re-legitimize a distinction between inside and outside, which for most of the other social transactions has lost any value. Hence, the traditional concept of the nation-state is reaffirmed and, through it, "a consistent self-definition of the state in times of globalization is forged" (Schinkel, 2009, p. 792). Schinkel, nevertheless, raises a crucial question: will this treatment of the problem of irregular migration prove effective, in the long run, in providing the state with new *raisons d'être*? What remains clear is that "Nation-states will not easily allot cosmopolitan rights (Habermas, 1993; Linklater, 1998), post-national (Soysal, 1994) or global citizenship (Dower, 2000) to irregular migrants, since precisely the creation of universal citizenship would entail providing the normative dimension of universal human rights with a legal dimension that necessarily compromises the traditional notion of the state" (Schinkel, 2009, p. 800).

The interdependence between the legal and the illegal, the regular and the irregular has been emphasized also in the works of Coutin. In her ethnography on Salvadorian irregular migrants, she has described their experience in terms of a permanent contradiction between presence and absence (Coutin, 2005a). Indeed, they are legally absent, since the authorities do not recognize them, yet, at the same time, they are physically present. In this sense, they perfectly embody Ngai's concept of "impossible subjects" (Ngai, 2014). As pointed out by Coutin: "although they 'cannot be', migrants continue to occupy the physical space. Their bodies therefore become a sort of absent space or vacancy, surrounded by law. The vacancies created by illegal presence make it possible for jurisdictions to remain whole" (Coutin, 2005a, p. 199). While the most common approaches explain irregular migration as a result of ineffective and powerless law, this perspective suggests an opposite understanding. "The law is not a force that bars illegal entry and sojourn; rather it is a process that defines who and what is illegal" (Coutin, 1996, p. 11; Garcés-Mascareñas, 2012, p. 31). In this sense, the construction of illegality is understood as a way for the state to establish and maintain the legal space against the illegal and "the regular nation", against "the irregular people".

These analyses offer indubitably interesting theoretical and conceptual understandings of irregular migration. The structural relationship between sovereignty and the exception that is proposed by Agamben sheds light on a similar interdependence between the state and the irregular migrant. To be sovereign, the state needs to decide on the exception, on what and who is inside or outside of the order that it creates. The irregular migrant is indeed the exception, the "bare life" against which the "political life" of the citizen acquires its significance. In this sense, her/his existence is vital to the existence of the very state. As put by Schinkel: "In the case of irregular migrants and their detention as 'illegal aliens' [...] the state tries to preserve a precarious balance between inclusion and exclusion, between *bios* and *zoe*" (Schinkel, 2009, p. 787). Nevertheless, while these conceptualizations can be helpful to understand the logic of sovereignty and its relation to the irregular migrant in abstract terms, it offers little explanation of the phenomenon in its concrete, sociological terms. On the one hand, the characterization of the irregular migrant as "bare life" or as an "impossible subject" and hence, as a completely excluded and subjugated victim of the state, is not matched by reality. Irregular migrants in many cases have rights and lead relatively normal lives. On the other, these interpretations do not offer clues to why the phenomenon assumes different forms and dimension within each national context.

Governmentality Techniques and Irregular Migration

An important strand of research, often inspired by the works of Michel Foucault, has interpreted irregular migration as a result of governmentality techniques enacted by states to better control their populations. From this perspective, the toleration of a certain degree of irregular migration, or the deliberate production of it by the state cannot be interpreted simply as a sovereignty imperative; instead, it needs to be considered as a "technology of power", as a legal and political construct aimed at effectively disciplining and managing populations (see for instance: Broeders, 2009; Chauvin & Garcés-Mascareñas, 2012; Chavez, 2007; Coutin, 2005a, 2005b; De Genova, 2002; Engbersen, 2001; Garcés-Mascareñas, 2012; Inda, 2006; Mezzadra, 2011; Morris, 2002; Rose & Miller, 1992; Thomas & Galemba, 2013; Vasta, 2011).

Foucault's concept of governmentality suggested a new understanding of power, one that surpassed the classic, top-down, coercive conception of it (Foucault, 1979, 2007, 2008). From his perspective, in order to be more effective, states have elaborated strategies to induce individuals to follow rules on the basis of their own will. This has been obtained through the development and use of a variety of new "technologies of power" that were meant to operate throughout the whole body of society. Schools, hospitals, psychiatric and penitentiary institutions, production sites and markets were the new sites where the state enacted its programmes and applied its strategies. The emergence of these new forms of power signalled exactly the switch from government to governmentality. The aim was no longer that of correcting single individuals through coercion, but of governing and disciplining the population as a whole through the induction of appropriate mentalities.

Within these new governmentality strategies, a crucial role was played by identification and surveillance technologies. In order for the modern states to apply their programmes, it was firstly necessary to build up administrative systems capable of identifying and classifying their populations. Yet, for Foucault, this step was not simply a functional requirement to accomplish other goals, but it was already a fundamental instrument of the new strategy. In *Discipline and punish: the birth of the prison*, he used Bentham's Panopticon as a metaphor to describe the functioning of the surveillance technique. In the disciplinary institution imagined by the English philosopher, prisoners could be seen at all times by guards who were invisible to them. The idea of being permanently observed induced them to behave according to the rules without the necessity to directly coerce them into doing so. According to Foucault, the effect of the Panopticon was: "..to induce in the inmate a state of conscious and permanent visibility that assures the automatic functioning of power. So to arrange things that the surveillance is permanent in its effects, even if it is discontinuous in its action; that the perfection of power should tend to render its actual exercise unnecessary; that this architectural apparatus should be a machine for creating and sustaining a power relation independent of the person who exercises it; in short, that the inmates should be caught up in a power situation of which they are themselves the bearers" (Foucault, 1979, p. 201). The extension of this strategy to the society as a whole was precisely the objective of the governmentality techniques: "on the whole, therefore, one can speak of the formation of a disciplinary society in this movement that stretches from the enclosed disciplines, a sort of social 'quarantine', to an indefinitely generalizable mechanism of 'panopticism'" (Foucault, 1979, p. 216).

It is within the context of this conception of society as a disciplinary system that the Foucauldian perspective has been related to management of migrant populations. The creation of different categories of migrants, to which different rights, duties, and limitations are assigned, would be a perfect example of a governmentality technique (Chavez, 2007; Inda, 2006; Vasta, 2011). The necessity to pass though the different categories and statuses before obtaining full citizenship would work as a "system of dams" (Mezzadra, 2011) that allows the selection of the appropriate candidates. The combination of this system of "civic stratification" (Morris, 2002) together with a powerful surveillance apparatus would induce migrants to enter a process of self-discipline, enabling power to work without having to exercise it. Each migrant knows that he/she is being observed, that by following the rules and fulfilling the requirements would take them ahead, while any fault or misconduct would take them back.

Within this system, the irregular migrant category plays a crucial role. Irregularity, rather than being a problem to be eliminated, has become for the state a fundamental component of the governmentality strategy. As put by Freise and Mezzadra: "Increasing mobility shapes the regimes of governmentality of the sovereign modern state and the ways in which power is distributed and enacted. Whereas historically state sovereignty was exercised through the control and surveillance of territory and subjects, governing no longer involves a delimited territory with spatially fixed and sedentary populations, but the control of highly mobile vagrant subjects and populations "menacing" the order and the security of states" (Friese & Mezzadra, 2010). Within this context: "The goal […] is not that of hermetically sealing off the borders of 'rich countries', but that of establishing a system of dams, of ultimately producing 'an active process of inclusion of migrant labour through its illegalization' (De Genova, 2002, p. 439). This entails a process of *differential inclusion* (Mezzadra & Neilson, 2010), in which irregularity emerges both as a produced condition and as a political stake in the politics of mobility" (Mezzadra, 2011, p. 229).

In a similar fashion, Garcés-Mascareñas has emphasized how the conception of irregular migration as an independent phenomenon from state policies and programmes is misleading: "While immigrant flows are indeed motivated by the importance of the structurally embedded demand for foreign workers in different receiving societies and of cross-national economic disparities and transnational economic, social and historical ties, these factors alone do not explain why a significant part of these flows take place illegally. The option (or the opportunity) to migrate legally or illegally cannot be understood without taking into account the obvious fact of the state and its migration policies. This is not only because it is the state that defines who may or may not enter, but because the state itself produces the migrants' illegality" (Garcés-Mascareñas, 2012, p. 205). From her perspective, more effective policies do not mean less irregular migration but more differentiated categories of migrants. In this sense, illegality does not function as an absolute marker of illegitimacy, but rather as a handicap within a continuum of "probatory citizenship" (Chauvin & Garcés-Mascareñas, 2012). The goal is not to completely exclude migrants but to make their inclusion more difficult.

The understanding of irregular migration as part of governmentality techniques employed by the state has evidenced another of its fundamental imperatives, i.e. that of managing populations. The role played by the law in establishing the conditions for formal inclusion and the related power to differentially (and conditionally) incorporate migrants into the host society certainly throws light on important aspects of the irregular migration phenomenon. In particular, they highlight the crucial role of the state in constructing the very category of the irregular migrant and the possible use of this power as part of its strategies to govern. However, these theories present a number of critical aspects. Whereas the status of irregularity is indubitably produced by the state and its creation may be in a way functional to the fulfilment of its interests, the phenomenon of irregularity, as regards its sociological dimensions, cannot be understood as a state product. The magnitude, characteristics and significance of irregular migration within a society can only be partly influenced by the state. In this sense, the concept of "production" is misleading, because it implies the producer's mastery over the process and the results that does not exist in this case. The distinction between the legal and sociological significance of a phenomenon becomes crucial. Even if the illegalization (or legalization) of a certain phenomenon is in the hands of the state, the social consequences of this are not. In this regard, the discussed theories tend to offer an image of the state, or more in general, of power, as rational, coherent, almighty forces that is not matched by reality. A state's action is fragmented, multi-levelled, sometimes contradictory and does not develop in a social void. Hence, the heterogeneity of forms, dimensions and characteristics that the irregular migration phenomenon displays in the different contexts in which it develops, can hardly be explained only as a governmentality strategy or as a technique of power. These power forces certainly exist and are employed by states but within a complex scenario of social actors and interactions.

Self-Restraint of States and Irregular Migration

As previously discussed, many scholars have interpreted irregular migration and the ineffectiveness of control policies as the result of a constraint over state capacities. This could be the result of external factors, as for instance, in the case of the effects of the international human rights regime, or of internal factors, as in the case of civil- society protests. Despite the different views on the causes of the constraints, these explanations have shared the idea of the state as a sort of "victim" or a passive recipient of them. In this section, the focus will move to the theories that have understood state limitations and, hence, phenomena such as irregular migration, not as the result of external restraints, but as the result of self-restraints.

The work of scholars, like Hollifield, Joppke and Guiraudon has focused on those internal characteristics of the contemporary states, and in particular, of the liberal-democratic ones, that determine a self-restraint in their ability to arbitrarily manage populations (Guiraudon, 2000; Guiraudon & Joppke, 2001; Guiraudon & Lahav, 2000; Hollifield, 1992, 2000, 2004; Joppke, 1998b, 1998c). For Hollifield, in order to fully comprehend the current difficulties of many states with regard to controlling migration, it is not enough to consider external economic, political or social factors. Instead, it is crucial to consider the role of endogenous political factors (Hollifield, 1992). In his analysis, the rise of rights-based politics in the US and Europe, after World War II, had a tremendous impact on state management of migration. This analysis did not underestimate the existence of a variety of influential actors and institutions demanding for more liberal policies towards migration, but argued that the extent and the ways in which their influence was possible depended on the inner structure and functioning of states. In liberal-democracies in particular, features such as constitutional charters, division of powers, judicial review of laws, and democratic representation were pinpointed as determinants in limiting the restrictive capacity of the system and in guaranteeing basic rights to everyone. The crucial point of the argument is that these features must not be considered as external and thus, somehow, as antagonists of the state. On the contrary, they should be considered as internal and thus consistent with the state's purpose. In this regard, Joppke has clearly stated: "..constitutional politics better explain the generosity and expansiveness of Western states towards immigrants than the vague reference to a global economy and an international human rights regime. The sovereignty of states regarding immigration control is more internally than externally restricted" (Joppke, 1998b, p. 20).

In a similar fashion, Guriaudon and Lahav have underlined: "Rather than global processes constraining domestic action, what we observe in the case of aliens' rights is a legally driven process of self-limited sovereignty. […] This means that the state has self-limited its capacity to dispose of aliens at will, once they have been admitted" (Guiraudon & Lahav, 2000, p. 189).

These analyses have offered another plausible explanation for the development of irregular migration. The phenomenon would appear not to be the result of states' failure or incapacity, but rather of states' application of rights-based liberalism. This form of self-restraint would seem to have severely limited the capacity of states to effectively deal with irregular migrants. For instance, practices, such as, mass expulsions, random identity checks or unjustified detention, just to mention the most important, that had been common features of migrant management, have become increasingly problematic. In addition, the existence of rights charters applicable to everyone and not just to citizens, and of an independent judiciary system has substantially empowered irregular migrants vis-à-vis states, allowing them to contest and, therefore, circumvent or delay the effects of their decisions and actions.

Two main critical considerations may be made about this claim. On the one hand, as shown for instance by Garcés-Mascareñas, also in non-democratic, non-liberal countries, where states have fewer limits to their restrictive capacities, irregular migration can be a sizable phenomenon (Garcés-Mascareñas, 2012). This fact evidences that, while internal political factors may certainly condition and make its development more difficult, irregular migration cannot be solely explained on the basis of these. In this respect, one may say that there are aspects of this phenomenon that escape political control (be they authoritarian or liberal), that exceed its capacities, and which are beyond its reach. On the other hand, this explanation seems to rely on the supposition that states are resolute in their opposition to irregular migration but that their internal functioning limits this determination. Many of the theories previously discussed have shown that this idea should at least be nuanced by considering that states may be interested in allowing certain levels of irregular migration.

3.3.2 *States and Social Demands*

An important set of theoretical explanations of irregular migration has departed from a very different concept of the role of states. Also in these approaches, it is the state that chooses to allow a certain level of irregular migration. The difference is that this choice does not respond to the state's own interests or imperatives, but to the demands coming from society.

Economic Interests and Irregular Migration

The explanation of irregular migration as a state product to fulfil the demands coming from the economic system has encountered enormous success in literature. The hypothesis is that "irregular migration serves to create and sustain a legally vulnerable, thus tractable and cheap, reserve of labour" (Garcés-Mascareñas, 2012, p. 29). This position has been developed in an impressive number of variants (Bach, 1978; Calavita, 1992; Castles, 2004; Castles & Kosack, 1973; Chavez, 1991, 2007; Cornelius & Rosenblum, 2005; Coutin, 2005a; De Genova, 2002, 2004, 2010; Goldring & Landolt, 2011; Köppe, 2003; Mezzadra, 2011; Piore, 1980; Portes, 1978; Portes & Bach, 1985; Samers, 2004; Sassen, 1988, 1996, 1998). Among these, it is possible to recognize different degrees of radicalism in the interpretation of the role of the state. Some accounts describe the state as a sort of puppet of capital; others offer a more nuanced view. Given the extent of this literature, the following discussion will be limited to a number of representative interpretations.

Marxist and segmented labour market theories provided a first interpretation that pictured states' ambiguity towards irregular migration as a strategy to meet the demand for cheap labour in industrialized countries (Bach, 1978; Castles & Kosack, 1973; Piore, 1980; Portes, 1978). In this view, the deep transformations that affected industrial economies during the 1960s and 1970s shaped a process of increasing segmentation of the labour markets. While native-workers, attracted by high-skilled, well-paid jobs, filled the upper part of the market, the lower part, consisting of precarious, unskilled and low-paid jobs, faced endemic shortages. States, then, combining labour and migration policies, were able to provide a stream of irregular foreign-workers. Their precarious status signified a flexibility and exploitability that was functional to the demands of the market.

Sassen has argued that deregulation and other policies furthering economic globalization cannot simply be considered as an instance of a declining significance of the state. On the contrary, deregulation and flexibilization must be considered as channels through which a growing number of states are furthering economic globalization and guaranteeing the right to global capital. Within this context, they continue to play a crucial role in the production of legality around new forms of economic activity. Moreover, with regard to the workforce, states are still decisive in generating the conditions for it to be available in the places, numbers and conditions required by producers. In this regard, the management of migration as a tool to differentially and conditionally include foreign workers, has become fundamental. This strategy is no longer simply a way to provide a "reserve army to overcome the periodical crisis of capitalism" (Castles & Kosack, 1973; Sassen, 1988); in fact, it has become a permanent structural mechanism within the new configuration of capitalism. These dynamic forces are particularly visible in global cities where, not just by chance, a "great concentration of corporate power and large concentration of 'others'" are discernible (Sassen, 1998). In a similar vein, Schierup, Hansen and Castles have underlined that: "Socially marginal migrants is not an imported phenomenon but rather 'part and parcel' of advanced capitalist strategies of deregulation, for the enhancement of 'flexibility' in terms of a networked economy and society, and a fragmented labour market" (Schierup et al., 2006, p. 299).

Departing from an analysis of historical and contemporary migration from Mexico to the Unites States, the works of De Genova, have argued against the tendency to "naturalize" migrants' "illegality" (De Genova, 2002, 2004, 2009, 2010), to treat it "as a mere fact of life, the presumably transparent consequence of unauthorized border crossing or some other violation of immigration law" (De Genova, 2004, p. 161). In his perspective: "..migrant 'illegality' signals a specifically *spatialized* socio-political condition. 'Illegality' is lived through a palpable sense of deportability – the possibility of deportation, which is to say, the possibility of being removed from the space of the US nation-state. The legal production of "illegality" provides an apparatus for sustaining Mexican migrants' vulnerability and tractability – as workers – whose labour-power, inasmuch as it is deportable, becomes an eminently disposable commodity. Deportability is decisive in the legal production of Mexican/migrant 'illegality' and the militarized policing of the US-Mexico border, however, only insofar as some are deported in order that most may ultimately remain (un-deported) – as workers, whose particular migrant status has been rendered 'illegal'" (De Genova, 2004, p. 161). The idea of a legal production of irregular migration on the part of states, which was previously discussed as part of a governmentality strategy, acquires here a more economic orientation. Migrants' vulnerability and tractability are created to provide the economic system with the docile and exploitable workforce it needs.

The main strength of these approaches has been to reveal the economic relevance of irregular migration in many contexts and to enquire into the political consequences of this. States, in these accounts, have been benevolent in according policies with the effect of generating important fluxes of irregular migrants, the kind of unskilled, flexible and exploitable workforce demanded by employers. While the question rose, namely the nexus between economy, politics and migration, is of great importance, the conclusion that states produce irregular migration to satisfy the demands of capital appears problematic. Firstly, this hypothesis does not explain why certain states are more resolute and efficient in fighting irregular migration than others (unless it is believed that in the first states capitalists have a higher morale), or why some decide to periodically regularize large numbers of migrants. More in general, it fails to account for the very differentiated picture of control policies and irregular migration realities, worldwide. Secondly, it presupposes that if states were not lenient to economic interests, they would be able to completely control irregular migration and this is quite unrealistic. Thirdly, it tends to downplay the existence of other interests, including those of the states themselves, which affect the formulation of policies. In this sense, the idea of the state as a weapon of capital is not convincing.

The State as a Broker of Social Demands: Pragmatic Solutions, Symbolic Policies

While the theories presented in the previous section relied on the hypothesis of the state being sensitive almost exclusively to economic interests, here the focus will be on the approaches that understand the state as a broker between a much wider set of social actors and interests. From this perspective, the magnitude and treatment of irregular migration within a certain context can be understood as the result of the pragmatic and, not always, transparent balances found by states between the different social demands and instances.

A very influential version of this interpretation has been proposed in the works of Freeman (Freeman, 1994, 1995, 2004, 2006). His political-economy model of policy-making aimed at explaining why in liberal-democracies, notwithstanding the widespread restrictionist rhetoric against migrations, the actual policies had been "broadly expansionist and inclusive" (Freeman, 1995). While the analysis acknowledged important differences within the analysed countries, basically related to the timing of their first experience of mass immigration and the extent to which migration policies are institutionalized, Freeman argued that the common tendency to expansive migration policies could be explained by the liberal and democratic characteristics of their political systems. The particular functioning of these systems implies that policies are the result of the interaction among three main players: (a) individual voters; (b) organized groups; (c) state actors. Nevertheless, when it comes to migration policies, what is Freeman's main argument is that the organized public dominates the bargaining. This is because immigration tends to produce concentrated benefits and diffuse costs, giving those who benefit from immigration greater incentives to organize than people who bear its costs. Hence, in this case, those who benefit, for instance, employers in labour-intensive industries and dependent on an unskilled workforce, businesses that profit from population growth (real estate, construction, etc.), and the family and ethnic relations of immigrants, have many more resources and capacities to make their voices heard than those who may be negatively affected by migration, namely the populations competing with immigrants for scarce jobs, housing, schools and government services. Since state actors are assumed to be vote-maximizers, they will respond to the organized pressure of groups favourable towards immigration, ignoring the widespread, but poorly articulated, opposition of the general public. The interactions will take place largely out of public view and with little outside interference. For these reasons, Freeman concludes that "[t]he typical mode of immigration politics is client politics" and client politics is strongly oriented towards expansive immigration policies. If this seems to be the general, long-term tendency, however, the politics of immigration in liberal democracies fluctuates and exhibits a tendency to go through predictable cycles. There is a 'good times/bad times' movement, in which migration is tolerated or encouraged during expansionary phases, but becomes the focus of anxieties when unemployment rises. While, in his works, Freeman has not explicitly intended irregular migration as a pragmatic solution to the conflict between the expansionary bias supported by organized groups, and the restrictive one supported at times by right-wing parties or portions of the public opinion, the idea has been undoubtedly suggested. For instance, in the conclusions of his article *Can Liberal States control Unwanted Migration?,* talking about how states deal with migration problems, he affirms that "states allow migration problems to accumulate and migration control policies to flounder until rising public pressure or some crisis makes action unavoidable" (Freeman, 1994, p. 30).

Zolberg has proposed the formula "wanted but not welcome" (Zolberg, 2000) to describe the existing tension in the majority of receiving societies: "...between two sets of concerns, represented as orthogonally related axes – the one representing economic interests, the other cultural and social interests, with a focus on 'identity'. Although migrants are highly prized on economic grounds, the massive internal migration or outright immigration of culturally distinct labour-market competitors triggers considerable uneasiness among the receivers on 'identitarian grounds'. [...] Because indigenous workers seldom have the power to prevent the immigration, it does take place; but the foreign workers are usually maintained in a state of segregation by way of an internal boundary. This facilitates their economic exploitation while minimizing their cultural impact (Zolberg, 2006, p. 225). The internal boundary to which Zolberg refers has historically acquired many different forms and degrees of impermeability. During the colonial era, for instance, this was constructed on racial grounds and put into effect in the slavery system. In contemporary societies, the internal boundaries are constructed in more subtle ways through the use of citizenship and identity policies. The case of irregular migration is precisely one where wanted migrants are kept from crossing the boundary on political bases.

The necessity on the part of states to reconcile the contrasting demands from society and the resort to pragmatic, often contradictory, solutions has led scholars to propose concepts such as "non-policy as a policy" or "symbolic policies" (Andreas, 1998; Broeders & Engbersen, 2007; Castles, 2004; Castles & Miller, 1993; Cornelius, 2005; Cornelius & Rosenblum, 2005; De Genova, 2004; Guiraudon & Joppke, 2001; McNevin, 2009; Schrover et al., 2008; Triandafyllidou, 2012). From this perspective, the combination of restrictive rhetoric and highly visible, but largely ineffective, policies is a pragmatic solution that allows states to show toughness and resolution against irregular migration without defrauding employers.

Cornelius, discussing the United States' migration policy, has spoken of a "manufactured" illegality. In his analysis, this results from a highly contradictory policy that combines border enforcing and legal-entry restrictions with weak internal controls. This "supply-side only" strategy cannot work. The "unrealistically low quotas for low-skilled foreign workers, quotas that are set so low for political rather than market-based reasons" (Cornelius, 2005, p. 789) implies a huge demand for irregular migrants. The enormous investments made by the government to reinforce the southern frontier, can only be interpreted as part of the "political calculus that heavy-handed, highly visible border enforcement remains useful in convincing the general public that politicians have not lost control over immigration" (Cornelius, 2005, p. 789).

Castles has explained the contradictions and apparent malfunctioning of migration policies, as the result of governments' difficulties in openly favouring one interest group and ignoring another. Therefore, a possible solution is the adoption of hidden agendas, i.e. "policies which purport to follow certain objectives, while actually doing the opposite". In particular, this regards migration policies, whose aim would be "to provide anti-immigration rhetoric while actually pursuing polices that lead to more immigration" (Castles, 2004, p. 214). Accordingly, Castles has underlined why it is important to consider that the declared objectives of states are often misleading, for instance, precisely regarding irregular migration: "Policies that claim to exclude undocumented workers may often really be about allowing them in through side doors and back doors, so that they can be more readily exploited. This, in turn, could be seen as an attempt to create a transnational working class, stratified not only by skill and ethnicity, but also by legal status" (Castles, 2004, p. 223). More recently, he has suggested that the contrast to irregular migration is a consensus instrument, vis-à-vis a tacit permissiveness (Castles, 2010a).

One of the strongest indications of the fictional character of the control efforts made by states has been found in their focalization on border controls. De Genova has spoken of the "border spectacle": "the spectacle of the enforcement of law at the border renders the racialized Mexican/migrant "illegality" visible, a "natural fact", whereas hides the production of that illegality" (De Genova, 2004, p. 177). Triandafyllidou has pointed out how fencing policies on their own, used without gate-keeping policies, are not effective but only spectacular (Triandafyllidou, 2010a, 2010b). Moreover, as shown by many scholars, while governments spend millions trying to stop irregular migrants from crossing their borders, the vast majority of them enter the countries legally with visas issued by the countries themselves (Finotelli, 2009; Finotelli & Sciortino, 2013; Morawska, 2001; Schrover et al., 2008; Sciortino, 2004a). To assess the effective functioning of control policies it is, therefore, necessary to look beyond the façade. As highlighted by Finotelli, the use of different control policies that may produce "unwanted" phenomena, such as, irregular migration, circulatory migration systems or the misuse of refugee policy, can be in effect a way for states "to handle the paradox between state control, market demands and the embedded liberalism of modern nation states" (Finotelli, 2009, p. 899).

The analysis of the different interests and social demands related to migration in each context to which states respond with particular pragmatic solutions, has led to questioning the idea of migration management as an undifferentiated practice worldwide. In fact, to get a better understanding of phenomena like irregular migration, it is necessary to consider: the social and political contexts within which they emerge (Finotelli, 2006, 2009), the different implementation cultures (Jordan et al., 2003), the relevance of the welfare provisions offered by states (Bommes & Geddes, 2000b; Bommes & Sciortino, 2011; Castles, 2004; Esping-Andersen, 1990; Schierup et al., 2006; Sciortino, 2004b; Williams & Windebank, 1998), the existence of certain administrative traditions or path-dependent processes (Faist, Gerdes, & Rieple, 2004; Finotelli, 2006; Finotelli & Echeverría, 2011; Van Der Leun, 2003).

The two main strengths of the theoretical approaches discussed in this section have been: (a) to present a much more complex view of the interests and actors related to migration in society; (b) to offer a differentiated picture of states' possible reactions to the social demands that may include pragmatic solutions, such as, the use of symbolic policies. Hence, the explanation of irregular migration becomes less straightforward than in other accounts and demands for a detailed analysis of the contexts and combinations of policies enacted by states. This approach opens up the path to a differentiated understanding of irregular migration, one that considers the diverse forms and dimensions the phenomenon acquires in each context, as well as the different relevance and significance it assumes. The critical points concern two aspects. On the one hand, there is a tendency in these perspectives to downplay states' own interests and picture them as more or less neutral brokers of the social interests. This tends to exclude the importance of the functional imperatives, such as, sovereignty or the control of populations, but also that of the administrative structures and of the different powers within the state. On the other, concepts like hidden agendas, symbolic policies or pragmatic solutions, at least in certain interpretations, seem to overstate both state capacity and rationality in governing migrations.

3.4 Critical Discussion of the Main Theoretical Explanations of Irregular Migration

The different works and approaches that have been analysed in this chapter have offered a wide number of different theoretical explanations of irregular migration. Each of them has added an important piece to the complex puzzle represented by contemporary migrations and, in particular, by the phenomenon of my concern. Nevertheless, each of them has also presented elements of criticism. In Table 3.2, all the approaches discussed, their logic, and the possible counterarguments, are summarized.

The conjunct analysis of these theories raises a striking problem. While, on their own, they provide persuasive elements to explain irregular migration, viewed together their claims are not always compatible and, in certain cases, are simply irreconcilable. Just to make one example, how can irregular migration be the result of a state strategy to control its population and, at the same time, the evidence that it has lost precisely that power? The problem, as many scholars have indicated, derives from a lack of theoretical ambition that has led to the production of case-specific, narrowly focused, unsystematic, hard to generalize explanations (Baldwin-Edwards, 2008; Bommes, 2012; Bommes & Sciortino, 2011; Cvajner & Sciortino, 2010). This has made the coexistence of contrasting hypotheses possible without the need to try to reconcile them. Therefore, the paradoxical situation of possessing a great number of theoretical explanations, each of which is able to illuminate a particular aspect of the phenomenon, but lacking a comprehensive theory capable of reconciling the many explanations and of offering a general interpretation was reached.

In this final section, an extensive discussion of the main critical aspects of the discussed theoretical explanations will be presented. This will lay the basis for the discussion of an alternative theoretical framework in Chap. 4.

3.4.1 Irregular Migration as an Undifferentiated, Mono-causal Phenomenon

Within the discussion of the single theoretical explanations of irregular migration, a number of critical aspects and possible counterarguments were raised, but here the focus will be on two main, wide-reaching problems that somehow drawn from all the others.

Firstly, there has been a general tendency to theoretically treat irregular migration as an undifferentiated phenomenon. This has led to underestimating the several, important differences the phenomenon has displayed in the various contexts in which it appeared, for instance, regarding its magnitude, characteristics, implications, etc. Yet, and more problematically, it has led to miscalculating the different causes at work in each circumstance. The main consequence has been the incautious

Table 3.2 Theoretical explanations of irregular migration

Explanation logic: irregular migration as the result of…				Counterarguments
State failure	Inherent limitations and weaknesses of states	Policy design	Knowledge production, policy design, predictive capacity limitations	States can be effective. If they are not it is because they do not want to.
		Policy implementation	Administrative, organizational and financial limitations	Irregular migration is not only a function of policies.
	External constraints and limitations	Economic globalization	The overwhelming force of the global economy	States have favoured globalization and its dynamics. Irregular migration is not a sign of their decline but of their choices.
		Political globalization	The role of: embedded liberalism; international legal and human rights' regimes; international institutions	States have the power to control; if they do not, this indicates possible collusions and self interests.
		Social globalization	Communications and transport technologies; information exchanges and cultural unification; transnational networks	Why do some countries control better than others?
		Migration industry (external part)	The activity of informal and criminal networks; human smuggling and human trafficking	Why differences between states?
				Why variation in the dynamics over time?
	Internal constraints and limitations	The informal economy	Informal employment in many production sectors	The states do not want to control the informal economy.
				No lineal relation informal economy – irregularity, the US case.
				Informal economy before irregular migration
		Migrants' agency	Individual strategies and counterstrategies to circumvent controls	Risk of overstating migrants' power and downplaying the role of structures.
				Why aspirations change?
		Internal social constraints	Street-level bureaucrats and other agents' discretionality	Policies are often effective.
			The role of civil society	Differences between countries.
			Migration industry (internal)	

(continued)

Table 3.2 (continued)

Explanation logic: irregular migration as the result of…				Counterarguments
State choice	Internal political factors	Sovereignty imperatives	State strategy to build its legitimacy and maintain sovereignty	Irregular migrants are not completely excluded. Sometimes they find it convenient to be irregular.
		Governmentality techniques	State strategy to control population	No differentiation.
		State self-restraint and rights-based liberalism	State self-constrained capacity to control populations	Irregular migration also exists in authoritarian states
				Irregular migration could be useful to states.
	The state and social demands	The state and capital	States produce irregular migrants to fulfil the demand of the labour market	States are not omnipotent
				Why do some states regularize?
				No differentiation
		The state as a broker between different social demands	Irregular migration as a pragmatic solution	States' own interests downplayed
			Controls as symbolic policies	States' capacities and rationality overstated

extension of the explicative logic that emerged for a specific case to other cases, or to "irregular migration" as an abstract concept. This problem, it was suggested, may have been related to a limited interchange between the empirical and the theoretical research on the field (Czaika & de Haas, 2011). While, since the 1990s, and especially down through the 2000s, a number of comparative analyses on irregular migration have increasingly evidenced the above-mentioned differentiated picture, their results rarely stimulated attempts to reconcile their implications and to produce a general theory.

Secondly, there has been a general inclination to offer mono-causal explanations. Not only was irregular migration explained as if it was the same phenomenon everywhere, but it was also described as if its causes could be reduced to one. Therefore, it was explained, for instance, as: the last bastion of sovereignty, a consequence of the informal economy, a by-product of Globalization, related to states' self-restraint, or the result of migrants' agency, just to mention a few. This tendency materialized in a paradigmatic manner in the dichotomy between the two main, broad, competing mono-causal explanations: that it was either the result of state failure or the result of state choices. Moreover, this approach has created the conditions for a limited debate about the different perspectives. If the dominant logic is the "either/or" one, the possibility for the "both/and" one is excluded.

These two problems can be related to a number of more specific and complex epistemological, conceptual and methodological ones. In the next sections some important critiques that have been advanced in the study of contemporary migrations more in general and that are significantly pertinent to our discussion will be examined.

3.4.2 *Epistemological Problems and Reductionisms*

The difficulties and contradictions that affect the interpretation of contemporary international migrations, which have become patent in our review, are probably an appendix to a more general difficulty in comprehending the epochal changes related to the so-called globalization. While these changes and the uncertainty they entail may be causing social concern, at least scientifically, they have favoured very interesting debates and critical analyses. In general, it may be stated that there has been a rising awareness of the complex challenges and questions posed by our time and of the necessity to improve our understanding. Many of these works offer important contributions to our discussion.

The Double-Edged Heritage of Methodological Nationalism

While the concept of "methodological nationalism" had been used before (Smith, 1983), it was successfully re-proposed within the globalization debate. Its success was probably related to the fact that it condensed into one concept the main critiques that had been moved to the so-called "mainstream literature". Wimmer and Glick-Schiller defined it as "the assumption that the nation/state/society is the natural social and political form of the modern world" (Wimmer & Glick-Schiller, 2002, p. 302). For many scholars, this assumption was one of the most important limitations to an adequate understanding of contemporary migrations (see, for instance: Beck, 2003; Friese & Mezzadra, 2010; Isin, 2009; Mezzadra, 2011; Schinkel, 2010; Wimmer & Glick-Schiller, 2002, 2003). Their criticisms demanded for an epistemological, conceptual and methodological turn within migration research, one that reconsidered the role of the state vis-à-vis that of other actors and, in particular, of migrants themselves.

Wimmer and Glick-Schiller identified three main variants of methodological nationalism: "1) Ignoring or disregarding the fundamental importance of nationalism for modern societies; this is often combined with 2) naturalization, *i.e.,* taking for granted that the boundaries of the nation-state delimit and define the unit of analysis; 3) territorial limitation which confines the study of social processes to the political and geographic boundaries of a particular nation-state. The three variants may intersect and mutually reinforce each other, forming a coherent epistemic structure, a self-reinforcing way of looking at and describing the social world" (Wimmer & Glick-Schiller, 2003, pp. 577–578).

In the works of Beck, the concept has been further developed. He indicated seven main principles:

> (a) The subordination of society to state, which implies b) that there is no singular, but only the plural of societies, and (c) a territorial notion of societies with state-constructed boundaries, i.e., the territorial state as a container of society. (d) There is a circular determination between state and society: the territorial nation-state is both the creator and guarantor of the individual citizenship rights and citizens organize themselves to influence and legitimate state actions. (e) Both states and societies are imagined and located within the dichotomy

> of national and international, which so far has been the foundation of the dominant ontology of politics and political theory. (f) The state as the guarantor of the social order provides the instrument and units for the collection of statistics about social and economic processes required by empirical social science. The categories of the state census are the main operational categories of empirical social science. [...] (g) In membership and statistical representation, methodological national operates on the either-or principle, excluding the possibility of both-and. But these oppositions – either "us" or "them", either "in" or "out" – do not capture the reality of blurring boundaries between political, moral and social communities... (Beck, 2003, pp. 454–455)

This perspective had the merit of questioning many well-established assumptions and of revealing the importance of phenomena, such as, transnationalism, migrants' activism or the emergence of new forms of citizenship. In particular, regarding the first aspect, the critique of methodological nationalism demanded for a return to the concept of society as the main conceptual tool for interpreting human relations. The state and the inventory of correlated concepts, such as, national population, national territory, sovereignty, citizenship must be considered as particular, historical constructs that permanently interact with others and constantly change.

As Wimmer and Glick-Schiller warned, however, an unbalanced criticism of methodological nationalism could entail its own risks: "..many who have attempted to escape the Charybdis of methodological nationalism are drifting towards the Scylla of methodological fluidism. It makes just as little sense to portray the immigrant as the marginal exception than it does to celebrate the transnational life of migrants as the prototype of human condition (Papastergiadis, 2000; Urry, 2000). Moreover, while it is important to push aside the blinders of methodological nationalism, it is just as important to remember the continued potency of nationalism" (Wimmer & Glick-Schiller, 2003, p. 600). This note of caution directly echoes back to our discussion of the already-mentioned problem of mono-causal explanations for irregular migration. While the critique of methodological nationalism has been crucial for revealing both state limitations and the key role of other actors at different levels, such as, migrants, global capitalism or international institutions, in many cases, this has led to a premature dismissal of any state relevance.

Reductionisms: The State and Society

As social sciences "originated in a "culture medium", politically and culturally framed by the nation state", this has determined a number of reductionist problems (Castles, 2010b). By this, what is meant is a tendency to analyse complex phenomena with simple, sometimes prejudicial, frameworks (Boswell, 2007).

A number of scholars, with arguments that often echo those opposing methodological nationalism, have further criticized the dominant conception of *the state-society relation*. Bommes has discussed the idea of the state as a "control unit of society" (Bommes, 2012, p. 166). For him, recalling the works of Luhmann, this idea entails a "limited concept of social structure" (Bommes, 2012, p. 20) that derives from the self-description of the state. He proposed considering nation-states, not as superposed, all-embracing containers of society, but as internal components of them. From this perspective, the analysis of irregular migration must, on the one hand, consider different capacities on the part of states, and on the other, their interaction with other actors within society. "Illegal migrations confront states with problems which draw attention to the necessities, possibilities and limitations of their migration policy… […]. Nation-states cannot renounce their right to control access to and residence in their territories. This right is implemented very differently in different states" (Bommes, 2012, p. 166). In certain societies, "the intervention of the state is wide-ranging and penetrates numerous areas of society" (Bommes, 2012, p. 166), while in others, it is lighter and more limited.

Hayman and Smart have warned against confusing "the legal" state and the "empirical" one (Heyman & Smart, 1999). While the former, the one envisaged in laws and in ideologies, denotes a solid, coherent, stable and socially-undisputed idea of it, the latter, the one emergent from reality, suggests a fragmented, complex, dynamic, wrangled actuality of it. In their analysis, it is precisely "illegal practices" that offer a privileged angle to disentangle the state-society equation and consider the relation as processual and conflictive.

Broeders signalled a tendency to fully believe, even within social sciences, in clichés constructed from a statist perspective (Broeders, 2009). For instance, powerful ideas, such as that of fortress or panopticon can be misleading, if not critically analysed. They "draw our attention to the power of the state and the enormous capacity it has built up in the 'fight against illegal immigration'" (Broeders, 2009, p. 37) but they may suggest that this power has become overwhelming or undisputed. While these ideas may well fulfil a social or political function or offer a clear, neat representation of social interactions, they may represent a problem for social sciences. Talking about surveillance and citing Bennett, Broeders pointed out: "Surveillance is, therefore, highly contingent. If social scientists are to get beyond totalizing metaphors and broad abstractions, it is absolutely necessary to understand these contingencies. Social and individual risk is governed by a complicated set of organizational, cultural, technological, political and legal factors" (Bennett, 2005, p. 133). Then he concludes by stating that, "This points to realities both inside and outside the power container of the state that are at odds with metaphoric clarity and lack of ambiguity" (Broeders, 2009, p. 37).

Another source of reductionism has regarded the internal conception of states. Many scholars have criticized the understanding of states as monolithic, coherent, and thoroughly rational actors (Boswell, 2007; Leerkes, 2009; Mezzadra, 2011; Van Der Leun, 2003). As expressed by Leerkes: "The state is not a monolithic whole either. Conflicts and different approaches and interests may emerge. There is a territorial division: municipal, provincial, national governments and supranational level (EU). There is power division: executive, judicial, legislative powers" (Leerkes, 2009, p. 29).

Boswell has detected two main tendencies, both problematic, in current accounts of state functioning as regards migration management. In the first, states have been characterized essentially as brokers of the different social interests. "The state is assumed as passively reacting to different interests. Its role is confined to that of finding a utility-maximizing compromise between organized interests" (Boswell, 2007, p. 79). In the second, states have been characterized as externally constrained in their ability to decide by: the other social actors, the liberal institution or the international society. From Boswell's perspective, both these tendencies lack "some understanding of the state's interests" and, in her opinion, the analysis of: "..its functional imperatives, must remain central to any political theory, especially one aiming to explain why under which conditions the state is constrained by liberal institutions. The state must remain central, since it continues to be the focus of expectation concerning the delivery of security, justice and prosperity" (Boswell, 2007, p. 88). She proposes four main, broad functional imperatives that characterize every state, in particular, they have to: (a) provide internal security for its subjects; (b) generate the condition for the accumulation of wealth; (c) provide a certain level of social "fairness"; (d) maintain institutional legitimacy. All these imperatives may be related to the migration phenomenon and are usually difficult to accomplish simultaneously. The crucial point in her analysis is that these are not considered predominant and disconnected from the rest of society; state actions and choices, constantly "resonate", in a mutually influencing relationship, with the rest of society.

A final problem related to the conception of the state and the understanding of migration relates to the treatment of states as if they were undifferentiated units. Many scholars have underlined the necessity to consider not only: a) the particularities of each state as regards their historical, institutional and political configuration, but also b) the particular way in which they give shape to a state-society relation in each context. Concerning the first aspect, criticisms were made towards a simplistic distinction between liberal-democratic and authoritarian states that often led to dichotomous abstract conclusions. The relationship between the political regime and migration, from this perspective, needs to be problematized and differentially analysed. An attempt in this direction was proposed by Rush and Martin who suggested different degrees of openness towards migration in relation to what they see as a trade-off between numbers and rights. The fewer rights that are guaranteed to migrants, the more numbers will be accepted and vice-versa (Ruhs & Martin, 2008). Garcés-Mascareñas's ground-breaking research was one of the first attempts to compare the management of migration by a liberal state with that of an authoritarian one. Her work pointed out that the relations between the political system and management of migration cannot be linearly interpreted (Garcés-Mascareñas, 2012). On the one hand, though it is certainly true that authoritarian states have fewer constraints in imposing their will over populations, this does not mean they are necessarily more effective in controlling migrant populations. On the other, the liberal-democratic character does not hold a state back from adopting ambiguous policies that, in many cases, more or less directly, entails the violation of its own constitutional principles. These results suggest the necessity for fully-fledged differential analyses that go beyond the labels and consider a number of factors, for instance: (a) the internal structure and functioning of each state; (b) the political culture and tradition; (c) the historical relation with migration; (d) the administrative and budgetary capacity of each state; (e) how policies are effectively implemented.

As for the second aspect, the state-society relations, an important strand of research raised attention on the necessity to differentiate the various forms and configurations of the welfare state (Bommes & Geddes, 2000b; Bommes & Sciortino, 2011; Esping-Andersen, 1990, 1996; Ferrera, 1996; Scharpf, 1996; Schierup et al., 2006; Sciortino, 2004b). Since the seminal work of Esping-Andersen, the study of the welfare state must not be intended as the simple distinction between the different rights and provisions offered by the administration to its citizens in each context (Esping-Andersen, 1990). Instead, his concept of welfare-state regimes pointed to the complex and dynamic interrelation between state activities, market characteristics and families' role in social provision. This approach led him to identify three main ideal-types of welfare regimes: the conservative/corporatist one, the liberal one and the social democratic one. Each regime implied the formation of a particular institutional framework and of a specific model of interaction with the other social structures. From this perspective, the role of the welfare-state regime becomes determinant in configuring, for instance, the employment structures and, thus, the new axes of social conflict and stratification.

The analysis proposed by Esping-Anderson set the conditions for a differentiated and more complex understanding of the relationship between welfare states and migrations. Bommes and Geddes departed from his work to reflect on this particular issue. For them, as they have clearly stated, "differentiation and specificity of argumentation is paramount": "Responses to immigration in national welfare states differed enormously with social inclusion and exclusion mediated by national historical, social and political contexts with a strong emphasis on territoriality and by diverse organizational and decisional infrastructures of different welfare state types. These are a major condition for the specific design of immigration and immigrant polices and have important consequences for the conditions of immigration, the status of migrants, and their social entitlements" (Bommes & Geddes, 2000c, p. 3). Therefore, they concluded that "it is the combination of specific national welfare types, their forms of inclusion and construction of the welfare community, their forms of immigration control and their ways of dealing with illegality" that finally shape the actual phenomenon of migration (Bommes & Geddes, 2000a, p. 253).

From a slightly different perspective, Devitt has underlined the necessity, in order to understand more fully contemporary migrations, to take into consideration the important differences existing among the "socio-economic regimes" in the receiving countries (Devitt, 2011). From her perspective, common explanations of migration determinants fail to account for the differentiated picture displayed, for instance, by European countries. Taking as a reference point, the attempts made in comparative capitalism literature (Deeg & Jackson, 2007; Jackson & Deeg, 2006) to cluster countries on the basis of the interlinking economic and industrial relations, employment, welfare, education and training regimes, she has proposed a "socio-economic institutional explanation for immigration variation in Europe". Her framework of analysis considers a number of variables and their direct or indirect effect on the demand for migrants. In particular, she has distinguished two main groups of variables, on the one hand, those specifically related to the job market: (a) the wage-skill of the economic regime; (b) the level of labour market regulation; (c) the employment-standards monitoring and, on the other, those related to what she calls the "surrounding system": (a) the welfare systems; (b) the education and training regimes; (c) the social services in relation to the demand for migrant care-workers (Devitt, 2011, pp. 587–591). In addition, Devitt highlights the need to consider the effects of the economic cycle, which may help to understand the inter-temporal variation within a single regime.

Reductionisms: Social Interactions

Another source of different forms of reductionism has come from the way in which social interactions have been (or have not been) understood. By social interactions, it is meant the way in which the different components of the social realm (actors, institutions, discourses, etc.) interact with each other and the effects that such interactions have. Two main critical and interrelated points that have been raised in this respect will be discussed.

Many scholars have mentioned, explicitly or implicitly, *the agency-structure relation* as a problematic aspect in the understanding of migrations (Boswell, 2007; Castles, 2010b; Van Meeteren, 2010; Van Nieuwenhuyze, 2009; Vasta, 2011). Boswell has referred to the debate concerning this issue in terms of a trade-off: "..between a theory with a plausible account of agency but which neglects social structures and one allowing substantial causal weight to institutions but lacking a plausible theory of agency" (Boswell, 2007, p. 76). She has related this trade-off to another, the one between theoretical neatness and complexity of explanation of social phenomena. The solution, as observed in the discussion of the different theories of irregular migration, has often been found in bypassing the problem and choosing to explain things, using either the structure prism or the agency one. The result has been a dichotomous tendency that has pictured irregular migrants either as "products" of structures (the state, the economy, the human rights' regime) or as a sign of their irrelevance.

More in general, structure and agency have been treated as alternatives, whereas the focus, as suggested by Vasta, should have been centred on their relation (Vasta, 2011, p. 3). In this respect, Van Nieuwenhuyze, recalling Giddens' structuration theory, has suggested: "Structure is not external to individual lives; structural properties are both the medium and the outcome of the practices they organize. Actions should be studied and analysed in their situated contexts, showing how they sustain and reproduce structural relations without falling into the functionalistic trap. There are no mechanical forces that guarantee the reproduction of a social system from day to day or from generation to generation, but all social life is generated in and through social praxis. In this sense, structure is internal, embodied; but it also stretches away in time and space, beyond the control of any individual actors. Through this approach, both structure and agency can be included in the analysis (Van Nieuwenhuyze, 2009, p. 16).

A second problematic aspect regards the understanding of *cause-effect relations* in society (Czaika & de Haas, 2011; Luhmann, 2012; Moeller, 2013). This issue has emerged especially in the analysis of state policies but it affects all social interactions. The problem concerns the epistemic structure of the input-output model that is usually used. The model postulates a direct, straightforward, exclusive relation between one event (the input or cause) and another (the output or effect) (Luhmann, 2012). While this model indubitably offers the advantage of theoretical neatness, its linear, one-dimensional structure generally fails to produce realistic accounts of social interactions. The model may work well for a simple laboratory experiment. There, it is possible to select a limited number of variables whose interaction needs to be studied (the internal variables) and to perfectly control them; other variables (the external ones) can be easily excluded. Furthermore, it is possible to delimit the length of the experiment. All these characteristics usually allow one to establish clear-cut, cause-effect relations. However, applied to social reality, this model tends to produce reductionist accounts. The complexity that characterizes the functioning of society poses different problems: (a) it is difficult to identify and control internal variables; (b) it is impossible to isolate the analysis from external variables; (c) it is not possible to temporally limit the interactions. The case of a new state policy offers a perfect example. The common understanding is that a certain action is enforced to obtain certain results, yet: (a) it is not possible to perfectly control both how the action is designed and implemented, and how the receivers adapt and counteract; (b) it is impossible to isolate external variables, for instance, the reaction of other social actors or the intervention of unconsidered factors; (c) it is not possible to temporally limit the effects generated by the initial action (for a similar discussion, see: Czaika & de Haas, 2011). Hence, it is very difficult to establish a straightforward cause-effect relation, at least in terms of the input-output model. The implication is not a negation of the existence of cause-effect relations, but the suggestion of an understanding of these as part of complex, multifactor, dynamic interactions.

The analysis carried out of the different theories of irregular migration has clearly showed the implications of this crucial epistemic problem in a number of tendencies: (a) the production of single-cause explanations; (b) the overstatement of actors' capabilities, rationality, vision; (c) in connection with the foregoing, the treatment of actors' actions or mis-actions (especially of institutional or system actors, e.g. "the state", "the economy", "society") in terms of intentionality; (d) the underestimation of external variables and other actors' reactions and forms of adaptation; (e) the understatement of the existence of short-term, medium-term and long-term effects. These tendencies have led to the construction of a straight cause-effect hypothesis about irregular migration. While these may have reached the goal of offering internally logical, clear-cut explanations, they have generally failed to offer comprehensive, generalizable theories capable of satisfactorily accounting for the complexity of the phenomenon.

The Sedentary Bias

The study of international migrations has also been affected by what has been defined as *the sedentary bias* (Bakewell, 2008; Castles, 2010b; Friese & Mezzadra, 2010; Papastergiadis, 2010; Zolberg, 2006). This tendency, which can be linked to methodological nationalism, interprets international migration as an exceptional phenomenon that perturbs the "normal" conditions of fixed national populations and limited cross-border fluxes. Since it frames them as exceptional, the adoption of this perspective treats migrations as a problem. As underlined by Zolberg: "Despite epochal changes, since nation-states emerged, they continue to adhere to the normative assumption that they consist of self-reproducing populations. In relation to this idea, emigration and immigration are constructed as disturbances" (Zolberg, 2006, p. 222). This perspective has oriented the scientific approach towards migrations in many ways, for instance: (a) human mobility has been understood as a problematic novelty instead of a normal, historical feature (Urry, 2007); (b) states and migrants have been interpreted as opposed, so the presence of the latter has then, somehow, indicated a failure of the former (Cornelius et al., 1994); (c) the internal demand for migrants has been neglected, leading to the invasion paradigm (Zolberg, 2006); (d) there has been a tendency to focus almost exclusively on the process of border trespassing, discarding, for example, the role of state visas (Bommes & Sciortino, 2011); (e) the state and its institutions, for instance, citizenship, have been considered as fixed and immutable, neglecting social interactions and change (Isin, 2009); (f) migrants have been treated as either victims or villains (Anderson, 2008); (g) the policy-oriented, problem-solving focus of research (Bommes & Sciortino, 2011). Hence, it is not difficult to recognize many of this problems lying behind the main explanations of irregular migration that have been discussed.

3.4.3 Summary: Problematic Aspects in the Theorizing of Irregular Migration

The analysis of the different theoretical explanations of irregular migration has revealed two main problems: the treatment of irregular migration as a undifferentiated phenomenon, both spatially and temporally, and the use of mono-causal explanations. These explanations, moreover, have been, at least in the way they have been proposed, difficult to reconcile, if not downright contradictory. The possibility for a more effective and comprehensive theory of irregular migration has been further limited by a number of theoretical problems and cul-de-sacs. To conclude, then, it seems possible to summarize a number of problematic points whose reformulation could help to develop a more adequate framework of analysis of irregular migration:

(a) There has been a problematic understanding of *society*. This has generally been understood as subsumed to the concept of the state. From this perspective, states were imagined not only as the containers of societies, but also as their regulators. The first aspect has favoured an undifferentiated analysis of irregular migration because every state has been assumed as an equal unit with similar characteristics, capacities, and functions. The second aspect, presupposing the possibility of total control, has led to "gap hypothesis"-like, failure/choice explanations. If it is assumed that states control society, a phenomenon that escapes their eye can only be interpreted either as a choice or as a failure.

(b) There has been a problematic understanding of the different *social actors* and their interests. These have generally been presented as rational, coherent and time-stable. The issue has been more glaring in the interpretation of institutional or systemic actors that possess a great degree of internal complexity, but it has also concerned the understanding of individuals. The tendency has been to paint them as single-minded, monolithic, steady actors, rather than as internally complex, often contradictory, interactive ones. In the explanations of irregular migration, the case of the state has been paradigmatic. Many approaches have proposed conceptualization, such as, state's desires, state's self-restraint or state's hidden agendas, which tend to produce reductionist interpretations. The state, like all the other social actors, is internally articulated, possesses different and often conflicting interests, interacts and adapts to the environment's stimulations.

(c) There has been a problematic understanding of *social interactions*. These have generally been understood in reductionist terms. On the one hand, this has determined a tendency to develop deterministic, direct cause-effect explanations of social interactions. Actors, policies, processes have been interpreted as perfectly capable of establishing and achieving their objectives, neglecting phenomena, such as, incoherence, ineffectiveness, or environmental reactions. On the other hand, there has been a tendency to produce dichotomous explanations, alternatively focused whether on the role of structures or of agency. Thus, irregular migration has been explained either as a phenomenon determined by the state, the economy or international law, or by the agency of actors, such as, migrants, smugglers or employers.

A theory that understands society as the main unit of analysis and the different actors, including the state, as internally complex, multifaceted, interactive ones, would probably offer the possibility for a more complex and differentiated theory of irregular migration. Moreover, it would allow the overcoming of the gap hypothesis conception. Once the idea is left behind that any actor or institution is internally monolithic and can control all social transactions, the whole focus changes. The query is no longer about actors' real intentions or covert plans, failure or success, domination or irrelevance; instead, it is about actors' decision processes and compromises, degrees of success or disappointment, complex and dynamic interactions. While this hermeneutic approach would certainly offer less deterministic and clear-cut accounts of irregular migration, its multi-causal and differentiated explanations would certainly attain the goal to be more congruous with social reality.

Bibliography

Agamben, G. (1998). *Homo sacer. il potere sovrano e la nuda vita*. Torino, Italy: Einaudi.

Agamben, G. (2003). *Stato di eccezione*. Torino, Italy: Bollati Boringhieri.

Anderson, B. (2008). *"Illegal immigrant": Victim or villain?* (Working Paper No. 64). Oxford, UK: Centre on Migration, Policy and Society.

Andersson, R. (2014). *Illegality, Inc.: Clandestine migration and the business of bordering Europe*. Berkeley, CA: University of California Press.

Andersson, R. (2016). Europe's failed 'fight' against irregular migration: Ethnographic notes on a counterproductive industry. *Journal of Ethnic and Migration Studies, 42*(7), 1055–1075. https://doi.org/10.1080/1369183X.2016.1139446

Andreas, P. (1998). The US immigration control offensive: Constructing an image of order on the southwest border. In *Crossings: Mexican immigration in interdisciplinary perspectives* (pp. 341–356). Cambridge, MA: Harvard University Press.

Bach, R. L. (1978). Mexican immigration and the American state. *International Migration Review, 12*, 536–558.

Bakewell, O. (2008). 'Keeping them in their place': The ambivalent relationship between development and migration in Africa. *Third World Quarterly, 29*(7), 1341–1358.

Baldwin-Edwards, M. (2008). Towards a theory of illegal migration: Historical and structural components. *Third World Quarterly, 29*(7), 1449–1459.

Baldwin-Edwards, M., & Kraler, A. (2009). *REGINE-Regularisations in Europe*. Amsterdam: Amsterdam University Press.

Balibar, E. (2001). *Nous, Citoyens d'Europe?* [We, citizens of Europe]. Paris: La Decouverte.

Balibar, E. (2010). At the borders of citizenship: A democracy in translation? *European Journal of Social Theory, 13*(3), 315–322. https://doi.org/10.1177/1368431010371751

Bean, F. D., Edmonston, B., & Passel, J. S. (1990). *Undocumented migration to the United States: IRCA and the experience of the 1980s* (Vol. 7). Washington DC: The Urban Insitute.

Beck, U. (2003). Toward a new critical theory with a cosmopolitan intent. *Constellations, 10*(4), 453–468.

Bennett, C. J. (2005). What happens when you book an airline ticket? The collection and processing of passenger data post 9/11. In E. Zureik & M. B. Salter (Eds.), *Global surveillance and policing. Borders, security, identity* (pp. 113–138). Portland, OR: Willan Publishing.

Biswas, S., & Nair, S. (2009). *International relations and states of exception: Margins, peripheries, and excluded bodies*. London/New York: Routledge.

Bommes, M. (2012). *Immigration and social systems: Collected essays of Michael Bommes*. Amsterdam: Amsterdam University Press.

Bommes, M., & Geddes, A. (2000a). Conclusion: Defining and redefining the community of legitimate welfare receivers. In M. Bommes & A. Geddes (Eds.), *Immigration and welfare. Challenging the borders of the welfare state* (pp. 248–253). London/New York: Routledge.

Bommes, M., & Geddes, A. (Eds.). (2000b). *Immigration and welfare. Challenging the borders of the welfare state*. London/New York: Routledge.

Bommes, M., & Geddes, A. (2000c). Introduction: Immigration and the welfare state. In M. Bommes & A. Geddes (Eds.), *Immigration and welfare. Challenging the borders of the welfare state* (pp. 1–12). London/New York: Routledge.

Bommes, M., & Kolb, H. (2002). Foggy social structures in a knowledge-based society–irregular migration, informal economy and the political system. *Unpublished Paper, University of Osnabruck*.

Bommes, M., & Sciortino, G. (Eds.). (2011). *Foggy social structures: Irregular migration, European labour markets and the welfare state*. Amsterdam: Amsterdam University Press.

Boswell, C. (2007). Theorizing migration policy: Is there a third way? *International Migration Review, 41*(1), 75–100.

Brochmann, G., & Hammar, T. (1999). *Mechanisms of immigration control: A comparative analysis of European regulation policies*. Oxford, UK/New York: Berg Pub Limited.

Broeders, D. (2009). *Breaking down anonymity digital surveillance on irregular migrants in Germany and the Netherlands*. Rotterdam, The Netherlands: Erasmus Universiteit.

Broeders, D., & Engbersen, G. (2007). The fight against illegal migration: Identification policies and immigrants' counterstrategies. *American Behavioral Scientist, 50*(12), 1592–1609. https://doi.org/10.1177/0002764207302470

Calavita, K. (1992). *Inside the state: The bracero program, immigration, and the I.N.S.* New York: Routledge.

Caplan, J., & Torpey, J. (2001). *Documenting individual identity: The development of state practices since the French Revolution*.

Castles, S. (2004). Why migration policies fail. *Ethnic and Racial Studies, 27*(2), 205–227. https://doi.org/10.1080/0141987042000177306

Castles, S. (2010a). Migración irregular: causas, tipos y dimensiones regionales. *Migración y Desarrollo, 8*(15), 49–80.

Castles, S. (2010b). Understanding global migration: A social transformation perspective. *Journal of Ethnic and Migration Studies, 36*(10), 1565–1586. https://doi.org/10.1080/1369183X.2010.489381

Castles, S., & Kosack, G. (1973). *Immigrant workers and class structure in Western Europe*. London/New York: Oxford University Press.

Castles, S., & Miller, M. J. (1993). *The age of migration. International population movements in the modem world.* New York.

Chauvin, S., & Garcés-Mascareñas, B. (2012). Beyond informal citizenship: The new moral economy of migrant illegality. *International Political Sociology, 6*(3), 241–259. https://doi.org/10.1111/j.1749-5687.2012.00162.x

Chavez, L. R. (1991). Outside the imagined community: Undocumented settlers and experiences of incorporation. *American Ethnologist, 18*(2), 257–278.

Chavez, L. R. (2007). The condition of illegality. *International Migration, 45*(3), 192–196.

Chiswick, B. R. (1988). Illegal immigration and immigration control. *The Journal of Economic Perspectives, 2*, 101–115.

Cornelius, W. A. (1982). Interviewing undocumented immigrants: Methodological reflections based on fieldwork in Mexico and the US. *International Migration Review, 16*, 378–411.

Cornelius, W. A. (2005). Controlling 'Unwanted' immigration: Lessons from the United States, 1993–2004. *Journal of Ethnic and Migration Studies, 31*(4), 775–794. https://doi.org/10.1080/13691830500110017

Cornelius, W. A., Martin, P. L., & Hollifield, J. (1994). *Controlling immigration: A global perspective*. Standford, CA: Stanford University Press.

Cornelius, W. A., & Rosenblum, M. R. (2005). Immigration and politics. *Annual Review of Political Science, 8*(1), 99–119. https://doi.org/10.1146/annurev.polisci.8.082103.104854

Cornelius, W. A., & Tsuda, T. (2004). Controlling immigration: The limits of government intervention. *Controlling Immigration: A Global Perspective, 3*, 7–15.

Coutin, S. B. (1996). Differences within accounts of US immigration law. *PoLAR: Political and Legal Anthropology Review, 19*(1), 11–20.

Coutin, S. B. (2005a). Being en route. *American Anthropologist, 107*(2), 195–206.

Coutin, S. B. (2005b). Contesting criminality: Illegal immigration and the spatialization of legality. *Theoretical Criminology, 9*(1), 5–33. https://doi.org/10.1177/1362480605046658

Cvajner, M., & Sciortino, G. (2010). Theorizing irregular migration: The control of spatial mobility in differentiated societies. *European Journal of Social Theory, 13*(3), 389–404.

Czaika, M., & de Haas, H. (2011). *The effectiveness of immigration policies. A conceptual review of empirical evidence* (Working Paper No. 33). Oxford, UK: International Immigration Institute.

De Genova, N. (2002). Migrant "Illegality" and deportability in everyday life. *Annual Review of Anthropology, 31*(1), 419–447.

De Genova, N. (2004). The legal production of Mexican/migrant "illegality". *Latino Studies, 2*(2), 160–185.

De Genova, N. (2009). Sovereign power and the "Bare Life" of Elvira Arellano. *Feminist Media Studies, 9*(2), 243–262.

De Genova, N. (2010). The deportation regime: Sovereignty, space and the freedom of movement. In N. De Genova & N. Peutz (Eds.), *The deportation regime: Sovereignty, space and the freedom of movement*. Durham, NC: Duke University Press.

de Haas, H. (2011). *The determinants of international migration*. IMI/DEMIG Working Paper. International Migration Institute, University of Oxford.

Deeg, R., & Jackson, G. (2007). Towards a more dynamic theory of capitalist variety. *Socio-Economic Review, 5*(1), 149–179.

Devitt, C. (2011). Varieties of capitalism, variation in labour immigration. *Journal of Ethnic and Migration Studies, 37*(4), 579–596. https://doi.org/10.1080/1369183X.2011.545273

Doomernik, J., & Jandl, M. (Eds.). (2008). *Modes of migration regulation and control in Europe*. Amsterdam: Amsterdam University Press.

Dower, N. (2000). The idea of global citizenship: A sympathetic assessment. *Global Society, 14*(4), 553–567.

Düvell, F. (2006). *Illegal immigration in Europe: Beyond control?* Basingstoke, UK/New York: Palgrave Macmillan.

Düvell, F. (2011). Irregular immigration, economics and politics. *CESifo DICE Report, 9*(3), 60–68.

Ellermann, A. (2010). Undocumented migrants and resistance in the liberal state. *Politics and Society, 38*(3), 408–429. https://doi.org/10.1177/0032329210373072

Ellermann, A. (2014). The rule of law and the right to stay: The moral claims of undocumented migrants. *Politics and Society, 42*(3), 293–308. https://doi.org/10.1177/0032329214543255

Engbersen, G. (2001). The unanticipated consequences of Panopticon Europe. In V. Guiraudon & C. Joppke (Eds.), *Controlling a new migration world* (pp. 222–246). London, New York: Routledge.

Engbersen, G., & Broeders, D. (2009). The state versus the Alien: Immigration control and strategies of irregular immigrants. *West European Politics, 32*(5), 867–885. https://doi.org/10.1080/01402380903064713

Engbersen, G., & Van Der Leun, J. (2001). The social construction of illegality and criminality. *European Journal on Criminal Policy and Research, 9*(1), 51–70.

Espenshade, T. J. (1995). Unauthorized immigration to the United States. *Annual Review of Sociology, 21*, 195–216.

Esping-Andersen, G. (1990). *The three worlds of welfare capitalism*. Princeton NJ: Princeton University Press.

Esping-Andersen, G. (1996). After the golden age? Welfare state dilemmas in a global economy. In G. Esping-Andersen (Ed.), *Welfare states in transition: National adaptations in global economies* (pp. 1–31). London: Sage.

Faist, T. (2000). Transnationalization in international migration: Implications for the study of citizenship and culture. *Ethnic and Racial Studies, 23*(2), 189–222.

Faist, T., Gerdes, J., & Rieple, B. (2004). Dual citizenship as a path-dependent process. *International Migration Review, 38*(3), 913–944.

Ferrera, M. (1996). The "Southern model" of welfare in social Europe. *Journal of European Social Policy, 6*(1), 17–37.

Finotelli, C. (2006). La inclusión de los inmigrantes no deseados en Alemania y en Italia: entre acción humanitaria y legitimación económica. *Circunstancia*, (10).

Finotelli, C. (2009). The north–south myth revised: A comparison of the Italian and German migration regimes. *West European Politics, 32*(5), 886–903. https://doi.org/10.1080/01402380903064747
Finotelli, C., & Echeverría, G. (2011). ¿Un mal país para vivir? Veinte años de inmigración en Italia. In L. Cachón (Ed.), *Inmigración y conflictos en Europa*. Barcelona, Spain: Hacer.
Finotelli, C., & Sciortino, G. (2013). Through the gates of the fortress: European visa policies and the limits of immigration control. *Perspectives on European Politics and Society, 14*(1), 80–101.
Foucault, M. (1979). *Discipline and punish: The birth of the prison*. New York: Vintage Books.
Foucault, M. (2007). *Security, territory, population: Lectures at the Collège de France, 1977–1978*. Basingstoke, UK/New York: Palgrave Macmillan: République Française.
Foucault, M. (2008). *The birth of biopolitics: Lectures at the Collège de France, 1978–1979*. Basingstoke, UK; New York: Palgrave Macmillan.
Freeman, G. P. (1994). Can liberal states control unwanted migration? *The Annals of the American Academy of Political and Social Science, 534*(1), 17–30.
Freeman, G. P. (1995). Modes of immigration politics in liberal democratic states. *International Migration Review, 29*(4), 881–902.
Freeman, G. P. (2004). Immigrant incorporation in western democracies. *International Migration Review, 38*(3), 945–969.
Freeman, G. P. (2006). National models, policy types, and the politics of immigration in liberal democracies. *West European Politics, 29*(2), 227–247. https://doi.org/10.1080/01402380500512585
Friese, H., & Mezzadra, S. (2010). Introduction. *European Journal of Social Theory, 13*(3), 299–313. https://doi.org/10.1177/1368431010371745
Garcés-Mascareñas, B. (2012). *Labour migration in Malaysia and Spain: Markets, citizenship and rights*. Amsterdam: Amsterdam University Press.
Geddes, A. (2001). Immigration and European integration: Towards fortress Europe? *Refugee Survey Quarterly, 20*(1), 229.
Giddens, A. (1990). *The consequences of modernity*. Cambridge, UK: Polity.
Glick-Schiller, N., Basch, L., & Blanc-Szanton, C. (1992). Towards a definition of transnationalism. *Annals of the New York Academy of Sciences, 645*(1), ix–xiv.
Glick-Schiller, N., Basch, L., & Szanton-Blanc, C. (1995). From immigrant to transmigrant: Theorizing transnational migration. *Anthropological Quarterly, 68*, 48–63.
Goldring, L., & Landolt, P. (2011). Caught in the work–citizenship matrix: The lasting effects of precarious legal status on work for Toronto immigrants. *Globalizations, 8*(3), 325–341. https://doi.org/10.1080/14747731.2011.576850
Guiraudon, V. (2000). The Marshallian triptych reordered: The role of courts and bureaucracies in furthering migrants' social rigths. In M. Bommes & A. Geddes (Eds.), *Immigration and welfare: Challenging the borders of the welfare state* (pp. 53–76). London/New York: Routledge.
Guiraudon, V., & Joppke, C. (Eds.). (2001). *Controlling a new migration world* (Vol. 4). London/New York: Routledge.
Guiraudon, V., & Lahav, G. (2000). A reappraisal of the state sovereignty debate: The case of migration control. *Comparative Political Studies, 33*(2), 163–195. https://doi.org/10.1177/0010414000033002001
Habermas, J. (1993). Citizenship and national identity: Some reflections on Europe. *Praxis International, 12*(1), 1–19.
Hanson, G. H. (2007). *The economic logic of illegal immigration*. New York: Council on Foreign Relations.
Heyman, J., & Smart, A. (1999). States and illegal practices: An overview. In J. Heyman (Ed.), *States and illegal practices*. Oxford, UK: Berg Publishers.
Hollifield, J. (1992). *Immigrants, markets, and states: The political economy of postwar Europe*. Cambridge, MA: Harvard University Press.

Hollifield, J. (2000). The politics of international migration. In C. Brettell & J. Hollifield (Eds.), *Migration theory: Talking across disciplines* (pp. 137–185). New York: Routledge.
Hollifield, J. (2004). The emerging migration state. *International Migration Review, 38*(3), 885–912.
Honig, B. (2009). *Democracy and the foreigner*. Princeton, NJ: Princeton University Press.
İçduygu, A. (2007). The politics of irregular migratory flows in the Mediterranean Basin: Economy, mobility and 'Illegality'. *Mediterranean Politics, 12*(2), 141–161. https://doi.org/10.1080/13629390701373945
Inda, J. X. (2006). *Targeting immigrants: Government, technology, and ethics*. Malden, MA: Blackwell Publishing.
Isin, E. F. (2009). Citizenship in flux: The figure of the activist citizen. *Subjectivity, 29*(1), 367–388.
Jackson, G., & Deeg, R. (2006). *How many varieties of capitalism? Comparing the comparative institutional analyses of capitalist diversity* (MPIfG Discussion Paper).
Jacobson, D. (1996). *Rights across borders: Immigration and the decline of citizenship*. Leiden, The Netherlands: Brill.
Joppke, C. (Ed.). (1998a). *Challenge to the nation-state: Immigration in Western Europe and the United States*. New York: Oxford University Press.
Joppke, C. (1998b). Immigration challenges to the Nation-State. In C. Joppke (Ed.), *Challenge to the nation-state: Immigration in Western Europe and the United States* (pp. 5–48). New York: Oxford University Press.
Joppke, C. (1998c). Why liberal states accept unwanted immigration. *World Politics, 50*(02), 266–293.
Jordan, B., Stråth, B., & Triandafyllidou, A. (2003). Comparing cultures of discretion. *Journal of Ethnic and Migration Studies, 29*(2), 373–395. https://doi.org/10.1080/1369183032000079648
Köppe, O. (2003). The leviathan of competitiveness: How and why do liberal states (not) accept unwanted immigration? *Journal of Ethnic and Migration Studies, 29*(3), 431–448.
Koser, K. (2010). Dimensions and dynamics of irregular migration. *Population, Space and Place, 16*, 181–193. https://doi.org/10.1002/psp.587
Kyle, D., & Koslowski, R. (2011). *Global human smuggling: Comparative perspectives*. Baltimore: JHU Press.
Lahav, G., & Guiraudon, V. (2006). Actors and venues in immigration control: Closing the gap between political demands and policy outcomes. *West European Politics, 29*(2), 201–223. https://doi.org/10.1080/01402380500512551
Leerkes, A. (2009). *Illegal residence and public safety in the Netherlands*. Amsterdam: Amsterdam University Press.
Leerkes, A., Van Der Leun, J., & Engbersen, G. (2012). *Crime among irregular immigrants and the influence of crimmigration processes* (pp. 267–288). Presented at the Social control and justice: Crimmigration in the Age of Fear.
Linklater, A. (1998). Cosmopolitan citizenship. *Citizenship Studies, 2*(1), 23–41.
López Sala, A. M. (2005). La inmigración irregular en la investigación sociológica. In D. Godenau & V. M. Zapata Hernández (Eds.), *La migración irregular una aproximación multidisciplinar* (pp. 161–180). Santa Cruz de Tenerife, Spain: Cabildo Insular de Tenerife, Área de Desarrollo Económico.
Luhmann, N. (2012). *Theory of society – Volume 1* (Vol. 1). Palo Alto, CA: Stanford University Press.
Mahler, S. J. (1995). *American dreaming: Immigrant life on the margins*. Princeton, NJ: Princeton University Press.
Martin, P., & Miller, M. (2000). *Employer sanctions: French, German and US experiences*. Geneva, Switzerland: ILO.
Massey, D. S. (1987). Understanding Mexican migration to the United States. *American Journal of Sociology, 92*, 1372–1403.
Massey, D. S. (1999). International migration at the dawn of the twenty-first century: The role of the state. *Population and Development Review, 25*(2), 303–322.

Massey, D. S., Arango, J., Hugo, G., Kouaouci, A., Pellegrino, A., & Taylor, J. E. (1998). *Worlds in motion. Understanding international migration at the end of the millennium* (pp. 1–59). Oxford, UK/New York: Clarendon Press.

Massey, D. S., Goldring, L., & Durand, J. (1994). Continuities in transnational migration: An analysis of nineteen Mexican communities. *American Journal of Sociology, 99*, 1492–1533.

McNevin, A. (2009). Contesting citizenship: Irregular migrants and strategic possibilities for political belonging. *New Political Science, 31*(2), 163–181. https://doi.org/10.1080/07393140902872278

Mezzadra, S. (2011). The gaze of autonomy. Capitalism, migration and social struggles. In V. Squire (Ed.), *The contested politics of mobility: Borderzones and irregularity* (pp. 121–143). London: Routledge.

Mezzadra, S., & Neilson, B. (2010). Borderscapes of differential inclusion: Subjectivity and struggles on the threshold of justice's excess. The borders of justice.

Moeller, H.-G. (2013). *Luhmann explained: From souls to systems*. Chicago: Open Court.

Morawska, E. (2001). Structuring migration: The case of Polish income-seeking travelers to the West. *Theory and Society, 30*(1), 47–80.

Morris, L. (2002). *Managing migration: Civic stratification and migrants' rights*. Abingdon, UK: Psychology Press.

Ngai, M. M. (2014). *Impossible subjects: Illegal aliens and the making of modern America*. Princeton, NJ: Princeton University Press.

Papadopoulos, D., Stephenson, N., & Tsianos, V. (2008). *Escape routes: Control and subversion in the twenty-first century*. London: Pluto Pr.

Papadopoulos, D., & Tsianos, V. (2007). *The autonomy of migration: The animals of undocumented mobility*. na.

Papastergiadis, N. (2010). Wars of mobility. *European Journal of Social Theory, 13*(3), 343–361. https://doi.org/10.1177/1368431010371756

Passel, J. S. (1986). Undocumented immigration. *The Annals of the American Academy of Political and Social Science, 487*(1), 181–200.

Passel, J. S., Cohn, D., & Gonzalez-Barrera, A. (2013). *Population decline of unauthorized immigrants stalls, may have reversed*. Washington, DC: Pew Hispanic Center. http://www.pewhispanic.org/files/2013/09/unauthorized-sept-2013-FINAL.pdf

Piore, M. J. (1980). *Birds of passage*. Cambridge Books.

Portes, A. (1978). Introduction: Toward a structural analysis of illegal (undocumented) immigration. *International Migration Review, 12*, 469–484.

Portes, A. (1996). Transnational communities: Their emergence and significance in the contemporary world-system. *Contributions in Economics and Economic History*, 151–168.

Portes, A. (2001). Introduction: The debates and significance of immigrant transnationalism. *Global Networks, 1*(3), 181–194.

Portes, A. (2003). Conclusion: Theoretical convergencies and empirical evidence in the study of immigrant transnationalism. *International Migration Review, 37*(3), 874–892.

Portes, A., & Bach, R. L. (1985). *Latin journey: Cuban and Mexican immigrants in the United States*. Berkeley, CA: University of California Press.

Portes, A., Guarnizo, L. E., & Landolt, P. (1999). The study of transnationalism: Pitfalls and promise of an emergent research field. *Ethnic and Racial Studies, 22*(2), 217–237.

Quassoli, F. (1999). Migrants in the Italian underground economy. *International Journal of Urban and Regional Research, 23*(2), 212–231.

Reyneri, E. (1998). The role of the underground economy in irregular migration to Italy: Cause or effect? *Journal of Ethnic and Migration Studies, 24*(2), 313–331.

Reyneri, E. (2004). Immigrants in a segmented and often undeclared labour market. *Journal of Modern Italian Studies, 9*(1), 71–93.

Rose, N., & Miller, P. (1992). Political power beyond the state: Problematics of government. *British Journal of Sociology, 43*, 173–205.

Ruhs, M., & Martin, P. (2008). Numbers vs. rights: Trade-Offs and guest worker programs. *International Migration Review, 42*(1), 249–265.

Samers, M. (2004). The 'underground economy', immigration and economic development in the European Union: An agnostic-skeptic perspective. *International Journal of Economic Development, 6*(2), 199–272.
Sassen, S. (1988). *The mobility of labor and capital: A study in international investment and labor flow*. Cambridge, UK; New York: Cambridge University Press.
Sassen, S. (1996). *Losing control? Sovereignty in an age of globalization*. New York: Columbia University Press.
Sassen, S. (1998). *Globalization and its discontents: Essays on the new mobility of people and money*. New York: New Press.
Scharpf, F. W. (1996). Negative and positive integration in the political economy of European welfare states. In *Governance in the European Union* (Vol. 15). London: Sage.
Schierup, C.-U., Hansen, P., & Castles, S. (2006). *Migration, citizenship, and the European welfare state: A European dilemma*. OUP Catalogue.
Schinkel, W. (2009). "Illegal Aliens" and the state, or: Bare bodies vs the zombie. *International Sociology, 24*(6), 779–806. https://doi.org/10.1177/0268580909343494
Schinkel, W. (2010). The virtualization of citizenship. *Critical Sociology, 36*(2), 265–283. https://doi.org/10.1177/0896920509357506
Schmitt, C. (2008). *The concept of the political: Expanded edition*. Chicago: University of Chicago Press.
Schneider, F., Buehn, A., & Montenegro, C. E. (2010). New estimates for the shadow economies all over the world. *International Economic Journal, 24*(4), 443–461.
Schrover, M., Van Der Leun, J., Lucassen, L., & Quispel, C. (2008). *Illegal migration and gender in a global and historical perspective*. Amsterdam: Amsterdam University Press.
Sciortino, G. (2004a). Between phantoms and necessary evils. Some critical points in the study of irregular migrations to Western Europe. *IMIS-Beiträge, 24*, 17–43.
Sciortino, G. (2004b). Immigration in a Mediterranean welfare state: The Italian experience in comparative perspective. *Journal of Comparative Policy Analysis: Research and Practice, 6*(2), 111–129.
Scott, J. C. (1998). *Seeing like a state: How certain schemes to improve the human condition have failed*. New Haven, CT: Yale University Press.
Scott, J. C. (2008). *Weapons of the weak: Everyday forms of peasant resistance*. New Haven, CT: Yale University Press.
Smith, A. D. (1983). Nationalism and classical social theory. *British Journal of Sociology, 34*, 19–38.
Soysal, Y. N. (1994). *Limits of citizenship: Migrants and postnational membership in Europe*. Chicago: University of Chicago Press.
Thomas, K., & Galemba, R. B. (2013). Illegal anthropology: An introduction. *PoLAR: Political and Legal Anthropology Review, 36*(2), 211–214.
Torpey, J. (1998). Coming and going: On the state monopolization of the legitimate "means of movement". *Sociological Theory, 16*(3), 239–259.
Triandafyllidou, A. (2009). Clandestino project final report. *Athen, 23*, 70.
Triandafyllidou, A. (2010a). *Controlling migration in southern Europe (Part 1): Fencing strategies*. ARI Instituto Real Elcano, 7.
Triandafyllidou, A. (2010b). *Controlling migration in southern Europe (Part 2): Gate-keeping strategies*. ARI Instituto Real Elcano, 8.
Triandafyllidou, A. (2012). *Irregular migration in Europe: Myths and realities*. Ashgate Publishing, Ltd.
Urry, J. (2000). Mobile sociology1. *British Journal of Sociology, 51*(1), 185–203.
Urry, J. (2007). *Mobilities*. Cambridge, UK: Polity.
Van Der Leun, J. (2003). *Looking for loopholes: Processes of incorporation of illegal immigrants in the Netherlands*. Amsterdam: Amsterdam University Press.

Van Meeteren, M. (2010). *Life without papers: Aspirations, incorporation and transnational activities of irregular migrants in the Low Countries*. Rotterdam, The Netherlands: Erasmus Universiteit.

Van Nieuwenhuyze, I. (2009). *Getting by in Europe's urban labour markets: Senegambian migrants' strategies for survival, documentation and mobility*. Amsterdam: Amsterdam University Press.

Vasta, E. (2011). Immigrants and the paper market: Borrowing, renting and buying identities. *Ethnic and Racial Studies, 34*(2), 187–206. https://doi.org/10.1080/01419870.2010.509443

Wallerstein, I. M. (2004). *World-systems analysis: An introduction*. Durham, NC and London: Duke University Press.

Williams, C. C., & Windebank, J. (1998). *Informal employment in the advanced economies: Implications for work and welfare*. Psychology Press.

Wimmer, A., & Glick-Schiller, N. (2002). Methodological nationalism and beyond: Nation–state building, migration and the social sciences. *Global Networks, 2*(4), 301–334.

Wimmer, A., & Glick-Schiller, N. (2003). Methodological nationalism, the social sciences, and the study of migration: An essay in historical epistemology. *International Migration Review, 37*(3), 576–610.

Zolberg, A. (2000). The politics of immigration policy: An externalist perspective. In *Immigration Research for a New Century* (pp. 60–68).

Zolberg, A. (2006). Managing a world on the move. *Population and Development Review, 32*(S1), 222–253.

Chapter 4
Understanding Irregular Migration Through a Social Systems Perspective

> *The structural change of society is beyond the observation and description of its contemporaries. Only after it has been completed and when it becomes practically irreversible, semantics takes on the task to describe what now becomes visible*
>
> Niklas Luhmann

The numerous problems and limitations that have affected the theoretical understanding of irregular migration cannot be simply related to a lack of empirical data or to the complexity of the phenomenon. Rather, they reveal important *obstacles épistémologiques* (Luhmann, 2007, p. 11) and conceptual problems that demand a reconsideration of many of the theoretical assumptions that have been generally used. Both the lack of differentiated analysis and the use of mono-causal explanations, which have been indicated as the most evident symptoms of theoretical ineffectiveness, have been linked to three broader and deeper causes. In particular: (a) a limited and often misguiding conception of *society*, usually subsumed within the concept of state; (b) the simplistic understanding of the different *social actors* and their interests; (c) the deterministic, cause-effect interpretation of *social interactions*. The extent and complexity of these issues, that evidently surpass the confines of the so-called migration studies, require a more general reflection on contemporary society and its functioning. From this perspective, international migration and, in our case, irregular migration, need to be considered as part and parcel, both products and determinants, of the broader social processes and structures. A satisfactory understanding of them can only be achieved in connection with a more general interpretation of contemporary society, one that critically reviews many important assumptions and preconceptions that have been imposed by the effects of methodological nationalism.

A particularly interesting and stimulating way to interpret international migrations in connection with the larger reflections of social theory has been attempted by a number of scholars who have applied Niklas Luhmann's social systems theory to the study of migrations. Following these steps and directly dealing Luhmann's

G. Echeverría, *Towards a Systemic Theory of Irregular Migration*, IMISCOE Research Series, https://doi.org/10.1007/978-3-030-40903-6_4

work, in this chapter some basic theoretical assumptions of his theory will be present and it will be suggested how they can offer alternative analytical tools to understand more adequately irregular migration as a structural and differentiated phenomenon of contemporary society.

4.1 The Semantics of the Modern State and Society

The extraordinary growth in the mobility of goods, capital, information and people, as well as the drastic reduction in the costs and time required for these exchanges, have shown the implausible character of the deeply-rooted understanding of society offered by the semantics of the modern state (Luhmann, 2009). In particular, globalization has helped to demonstrate the questionability of one of the central assumption of that semantics: the idea of politics as a preeminent, overarching force, capable of fully embracing and controlling society (Luhmann, 2009, p. 79). The analysis of the contradictions between that semantics and what emerges from the structural reality is, therefore, a fundamental step in order to develop an alternative understanding.

An abstract representation of contemporary global society, one that does justice to the myriad of exchanges that take place worldwide, would probably appear as a complex web of lines and colours that mix in exceedingly intricate ways. The image would represent both the diversified communications that interact and connect in seemingly random and disparate ways in every corner of the world and the variety of population encounters, migrations, and contaminations that implicate all ethnic groups, cultures, religions and traditions. Clear demarcations, unique identities, original peoples, if they ever existed outside political discourses, would be impossible to locate. It is possible that a painting by Jackson Pollock could offer a good visual approximation of such a society. It would appear as a largely unified, global space of interaction (Fig. 4.1).

Yet, if we had to graphically imagine the conceptualization of society proposed by the semantics of modern politics, we would come up with a completely different picture. A painting by Piet Mondrian could probably offer an excellent approximation. Black neat lines would perfectly separate a number of internally-homogeneous areas, and the result would look somewhat similar to that of ordinary political maps. The "social space", understood as the space where social transactions take place, would fall entirely within the "political space", understood as the space where those transactions are regulated and legitimized by a sovereign power. Accordingly, society would not appear as one, but as many societies, each corresponding to a single state and its own well-demarcated territory. In this idealization, the political power, embodied by the state and its institutions, since it is able to regulate all social transactions, becomes, at the same time, the enforcer and "the guarantor of the social order" (Luhmann, 2009, p. 79). To make this possible, a crucial step is to define a particular population and to be able to effectively distinguish between those people

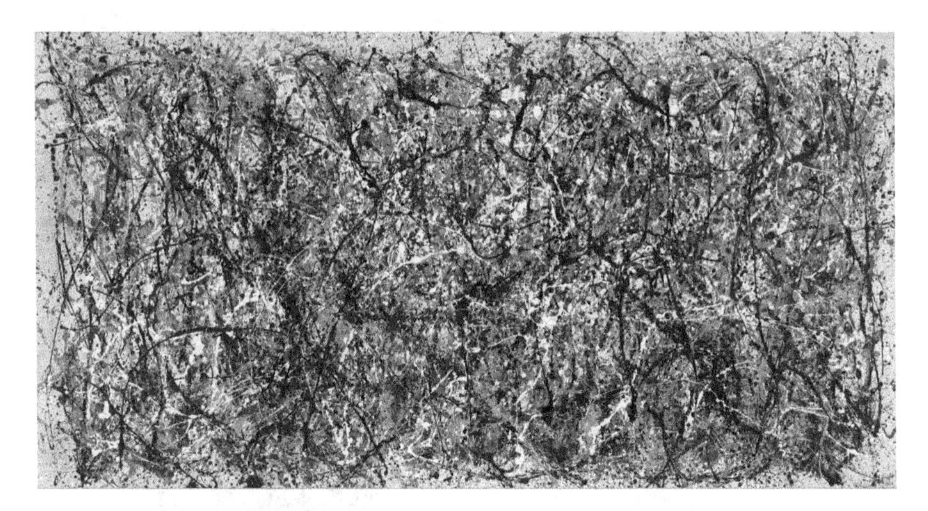

Fig. 4.1 Jackson Pollock, *Convergence* (1952)

considered insiders, the citizens, and the outsiders, the foreigners. The modern nation-state accomplishes this task precisely through the concept of "the nation" which establishes a natural, direct and unbreakable link between each individual (the population), the place of birth (the territory), and the political power over that territory (the state) (Luhmann, 2009, pp. 227–236; Schinkel, 2010). As becomes evident, in this representation, the concept of society is subsumed into the one of the state: in order to participate in the former, it is necessary to be part of the latter; in order to participate in social transactions, it is necessary to be citizens (Fig. 4.2).

As one can observe, the two paintings offer a completely different interpretation of society. This implies that, while the semantics of modern politics has certainly dominated the modern understanding of society and politics, serving as the ideological pillar for the affirmation of the modern nation-state as the main form of political organization worldwide, its ideals have never been fully realized (Luhmann, 2009, p. 85). This fact, which today is starting to appear self-evident, was not so obvious just some years ago. In the previous historical phase, thanks to the affirmations of modern politics, the "social space" tended to overlap more with the "political space", giving the impression that the "Mondrian world" was plausible. With the rise and development of the nation-state, the majority of social interactions were increasingly restricted within the national boundaries and those that crossed frontiers were rather limited and closely controlled. This tendency also affected human mobility. Throughout the nineteenth century and especially after the First World War, migrations were heavily restricted and, when they did occur, they were done through the channels established by the states and often under their own auspice (see Chap. 2).

However, even if this historical phase certainly favoured a growth in the political capacity to intervene and regulate social transactions, the world imagined by the semantics of the modern state never materialized. Even during the apex of the

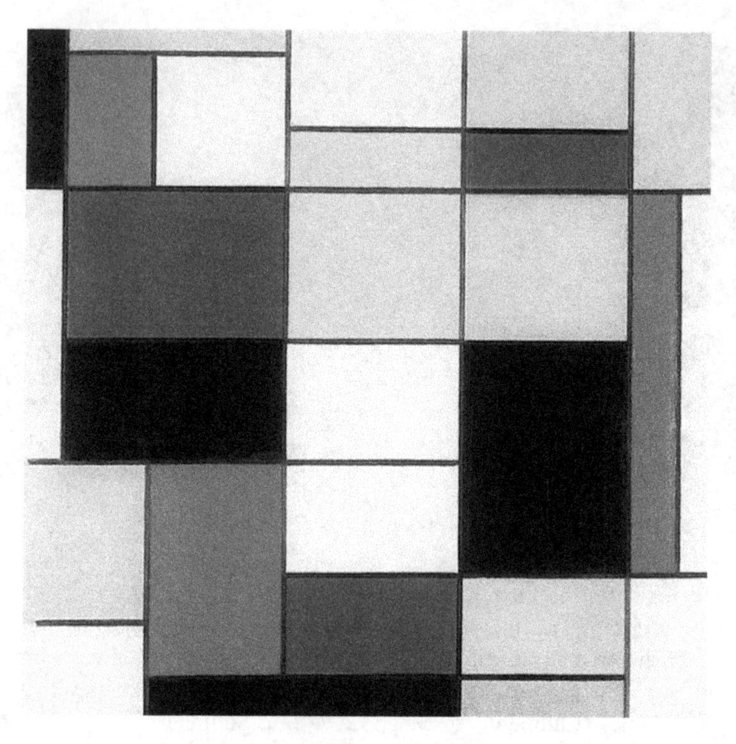

Fig. 4.2 Piet Mondrian, *Composition A* (1923)

mercantilist ideas, the economic exchanges beyond the limits of the state continued to take place (Luhmann, 2009, p. 85). At the same time, notwithstanding the dream (or the nightmare) of a sedentary world (see Chap. 3), migrations never disappeared. After the Second World War, the development of interconnections between individuals and groups gained new strength, inverting that overlapping movement between the "social space" and the "political space". In this sense, globalization has been determining a "spill over" of the "social space" beyond the boundaries of the "political space" as was prefigured by the modern state. As pointed out by Schinkel, in reference to migrations, if maybe: "for a brief ('Marshallian') period in the 20th century, citizenship sufficed as a guarantor of membership of both nation-state and the discursive domain of society in an age in which flows of migration have become permanent, that is no longer plausible. [...] The moment society is entered by people not tied through nativity to the nation, the nation can no longer be seen to overlap relatively with society" (Schinkel, 2010, p. 267).

For Luhmann, the incongruence between the two representations of society shows:

> A typical case of lack of synchronization between the structure of society and semantics. While in other fields of society – for instance, in the intimacy relations – the ideological baggage of semantics produced transformations that profoundly affected social structures. In the field of politics what we observe instead is the maintenance of a conceptual

> framework that has been overwhelmed. The problem is [...] that the semantics of politics takes surreptitiously the role that should correspond to the concept of society[1] (Luhmann, 2009, p. 79).

This lack of synchronization between structures and semantics implies that, while the incongruences between the structural reality and semantics are becoming uncontestably evident, the method to interpret them is still deeply influenced by semantics and its conceptualizations. As was widely discussed in Chap. 3, this produces a number of theoretical problems, which directly affect our ability to understand a phenomenon like irregular migration. In the next section, it will be discussed how social systems theory, departing precisely from a reformulation of the concept of society that re-establishes its central position, can offer a theoretical framework capable of avoiding many of the problems mentioned.

4.2 Elements of Niklas Luhmann's Social Systems Theory

Luhmann's social systems theory is extremely complex and ambitious. The explicit attempt by the German sociologist was nothing less than to build a comprehensive "theory of society" (Luhmann, 2012). This project which was accomplished throughout a lifetime research, transformed into a monumental effort to analyse and re-define many consolidated conceptions and ideas. Given the complexity and the extension of his work, in the next sections there will not be and attempted to summarize his theory. Vice versa, some of its concepts and ideas will be presented and it will be discussed how they can be useful to develop a better understanding of irregular migration.

4.2.1 Systems

The fundamental concept at the root of social systems theory is precisely the one of system. Luhmann proposes a very general and abstract definition: "a form with two sides"; a form that creates a difference, "a difference between system and environment" (Luhmann, 2006, p. 45, 2012). He considers three main kinds of systems: living systems (cells, organisms), psychic systems (minds) and social systems (function systems, organizations, interactions). All these systems share two crucial characteristics: they are autopoietic and operationally closed.

With the first term, i.e. autopoiesis, mutated from biology, Luhmann means that every system creates itself as a chain of operations in a process of circular self-production.

[1] The translation from Spanish is mine.

> Autopoietic systems are systems that themselves produce not only their structures but also the elements of which they consist in the network of these same elements. The elements (which from a temporal point of view are operations) that constitute autopoietic systems have no independent existence. They do not simply come together. They are no simply connected. It is only in the systems that they are produced (on whatever energy and material basis) by being *made use of as distinctions* (Luhmann, 2012, p. 32).

The concept of autopoiesis implies that:

> …all explanations start with the specific operations that reproduce a system". In this sense the concept "says nothing about what specific structures develop in such system […]. Nor does it explain the historical states of the system from which further autopoiesis proceeds. […] Autopoiesis is therefore not to be understood as the production of a certain "gestalt" [form]. What is decisive is the production of a difference between system and environment (Luhmann, 2012, pp. 32–33).

With the second term, i.e. operational closure, Luhmann defines the way in which systems relate to their environment.

> "There is no input of elements into the system and no outputs of elements from the system. The system is autonomous, not only at the structural level, but also at the operational level. This is what autopoiesis mean. The system can constitute operations of its own only further to operations of its own and in anticipation of further operations of the same system" (Luhmann, 2012, p. 33). "At the level of system's own operations there is no ingress to the environment, and environmental systems are just as little able to take part in the autopoietic processes of an operationally closed system" (Luhmann, 2012, p. 49).

In other words, the relation between a system and its environment cannot be interpreted with an input/output model. Elements or events become relevant for a system only as they transit through the channels and mechanisms built by the system to observe its environment. Through this process of filtering and re-assembling, systems construct their own "systemic reality".

As pointed out by Moeller, this conceptualization produces a radical shift from the common understandings of reality:

> The theory of autopoiesis and operational closure […] breaks with the notion of a common reality that is somehow "represented" within all systems or elements that take part in reality. According to systems theory, systems exist by way of operational closure and this means that they each construct themselves and their own realities. How a system is real depends on its own self-production, and how it perceives the reality of its own environment also depends on its self-production. By constructing itself as a system, a system also constructs its understanding of the environment. And thus a systemic world cannot suppose any singular, common environment for all systems that can somehow be "represented" within any system. Every system exists by differentiation and thus is different from other systems and has a different environment. Reality becomes a multitude of system-environment constructions that in each case are unique (Moeller, 2013, p. 16).

Autopoiesis and operative closure do not mean absolute closure. All systems relate to their environment and in this sense they are open, yet not operationally open. This means that the environment cannot *directly* affect the internal functioning of a system, i.e. its internal operations. The input/output model cannot be of help for understanding systemic relations, because it presupposes the possibility of an *immediate* contact of the environment with a system and of a system with the environment (and other systems). Social systems theory, instead, understands these relations as *mediated* by the *ad hoc* cognition structures and mechanisms that each system develops to relate with the outside. Elements, events, irritations present in the environment become relevant for a system only if they are successfully translated into its internal language, becoming information. "Such information does not exist in the environment but only has correlates out there…[…]" (Luhmann, 2002, p. 122 in Moeller, 2013, p. 17). What a system sees through its mechanisms, what a system makes of the irritation it receives, is entirely dependent on its own structure (Moeller, 2013, p. 17). This strategy allows systems to reduce the complexity present in their environment and, therefore, to be able to build up their own internal systemic complexity. Moeller provides illustrative examples:

> A system cannot come into immediate contact with its environment by way of its own operations. The biological operations within a cell, for instance, are only connected to and in continuation with the other biological operation within it. The same is true for psychic operations within an individual mind and for communicational operations within a communication system. The biological operations of the brain are connected to and continued by other biological operations of the brain. Similarly, a thought or a feeling is connected to and continued by other thoughts or feelings. A mind cannot continue a thought with a brainwave. And a communication can, of course, only be continued with more communication. You cannot communicate with me with your mind or brain, you will have to perform another communicative operation such as writing or speaking (Moeller, 2013, p. 17).

While social systems theory excludes the possibility of direct interaction between systems, the concept of *structural coupling* captures the possibility of a strong interdependence. Two systems are structurally coupled whenever the presence of the other one in each environment is so "bulky" that the structures on which the autopoiesis rely become shared. The operative closure is preserved since the coupling "only affects the structures level and not that of self-reproduction: while systems' independence remains intact in what refers to the construction of their own elements and the determination of their contacts, it is possible to observe a coordination between reciprocal structures" (Baraldi, Corsi, & Esposito, 1996, pp. 19–21).

4.2.2 Social Systems and Society

Social systems are a specific kind of system defined by their distinctive operation: communication (Luhmann, 2009, p. 91, 2012, p. 41). Biologic systems and psychic systems are the environment of social systems. Communication can be "made" by means of a wide variety of communicational elements, for instance: gestures, images, sounds, languages, money, etc. In order for one of these elements to become communication, and not simply be a body movement, a visual object, a noise, a group of signs, or a piece of paper, it must be inserted into a sequence that makes it possible to overcome the double-contingency problem and therefore produce understanding (Luhmann, 2009, p. 645). The autopoietic development of different types of sequences that use different types of communication elements produces a wide variety of social systems.

Society is defined by Luhmann as the "all-comprehensive social system" (Luhmann, 2012, p. 40). In their "Luhmann Glossary", Baraldi, Corsi and Esposito effectively summarize the sociologist's conceptualization:

> Society is a special type of social system; the social system that includes all communications. As a consequence there is no communication outside society. […] All differentiation of particular social systems occurs within society. Society, intended as system, is not made of individuals, their relations or roles, is made of communications. The boundaries of society are not the territorial ones, but those of communication. […] The distinctiveness of society as a social system relies on its complexity reduction achievement: society is the social system that institutionalizes the latest, most basic complexity reductions and, through this, creates the premises for the operations of the other social systems[2] (Baraldi et al., 1996, pp. 154–155).

This complex all-embracing system is internally diversified into a wide variety of sub-systems. Each sub-system, which performs a specific type of communication, has its own autopoietic independence and is operationally closed. As pointed out by Moeller, this last point implies that every subsystem is the intra-social environment for the others. In this sense, each one "has its own social perspective and creates its own reality. […] Society looks different from the perspective of each subsystem and there is no perspective, or super-system that can "supervise" the subsystems" (Moeller, 2013, p. 24) (Fig. 4.3).

As for systems in general, including social systems, operative closure determines an indirect, mediatory form of interaction with the environment and the other systems (both intra-social and extra-social). Luhmann uses the concepts of *irritation* and *resonance* to specify more clearly the ways in which this interaction takes place (Luhmann, 1990, p. 61, 2012, p. 67). If one social system, as part of its own autopoietic process and by means of its own structures, emits a communication to the environment, this communication has the effect of irritating the other systems. This external irritation is filtrated and translated by the observing structures of the receiving system into its

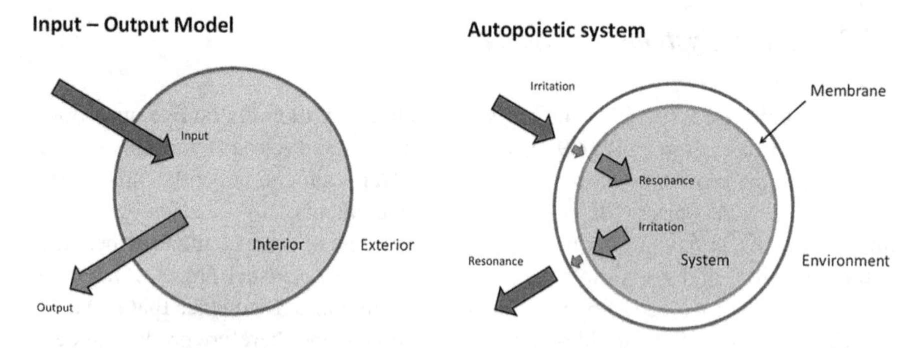

Fig. 4.3 Input-Output model vs. Autopoietic systems' model

[2] The translation from Spanish is mine.

internal operations. This may or may not produce a systemic resonance, understood as a reaction that is entirely dependent on the specific structures and characteristics of this system. To give one example, if the political system takes a political decision, for instance, to raise taxes, this has the effect of irritating the other systems in its environment. Each system perceives this irritation in a particular way and reacts or, rather, resonates according to its own internal logic. In this case, the economic system may resonate by raising prices, the mass-media system by airing protests, the legal system by signalling the unconstitutionality of the measure, etc. The crucial point is that no system can directly interfere with or determine the operations of the others or of the entire society. The irritation/resonance model bestows systemic independence on each system and understands interactions as processes of indirect, mutual influence. As pointed out by Moeller:

> Through structural coupling, systems cannot steer other systems or directly interfere in their operation. They can, however, establish relatively stable links of irritation that force other systems to resonate with them. There are always two sides to structural coupling. A system that irritates another cannot, in turn, avoid being irritated (Moeller, 2013, p. 39).

4.2.3 Social Differentiation and Modern Society

Society, the all-comprehensive social system, is internally differentiated (Luhmann, 2013, pp. 1–16). In Luhmann's opinion, systemic differentiation cannot be understood through the whole/parts scheme.

> It is important to understand this process with the necessary precision. It does not involve the decomposition of a 'whole' into 'parts', in either the conceptual sense (*divisio*) or the sense of actual division (partition). The whole/part schema comes from the old European tradition, and if applied in this context would miss the decisive point. System differentiation does not mean that the whole is divided into parts and, seen on this level, then consists only of the parts and the 'relation' between the parts. It is rather that every subsystem reconstructs the comprehensive system to which it belongs and which it contributes to forming through its own (subsystem-specific) difference between system and environment. Through system differentiation, the system multiplies itself, so to speak, within itself through ever-new distinctions between systems and environment in the system. The differentiation process can set in spontaneously; it is a result of evolution, which can use opportunities to launch structural changes. It requires no coordination by the overall system such as the schema of the whole and its parts had suggested. [...] The consequence is a differentiation of societal system and interaction systems that varies with the differentiation form of society (Luhmann, 2013, p. 3).

Luhmann sees the particular form of social differentiation as the result of social evolution (Luhmann, 2009, pp. 380–384). As pointed out by Baraldi et al.:

> What evolutionarily varies and measures social evolution is the form of primary differentiation. This form establishes the structure of society: social evolution consists in mutation of the social structure. Society primarily differentiates into partial sub-systems that produce more restricted communications. [...] These partial systems do not need to distinguish communication from what is not communication, since for that it is enough for them to be part

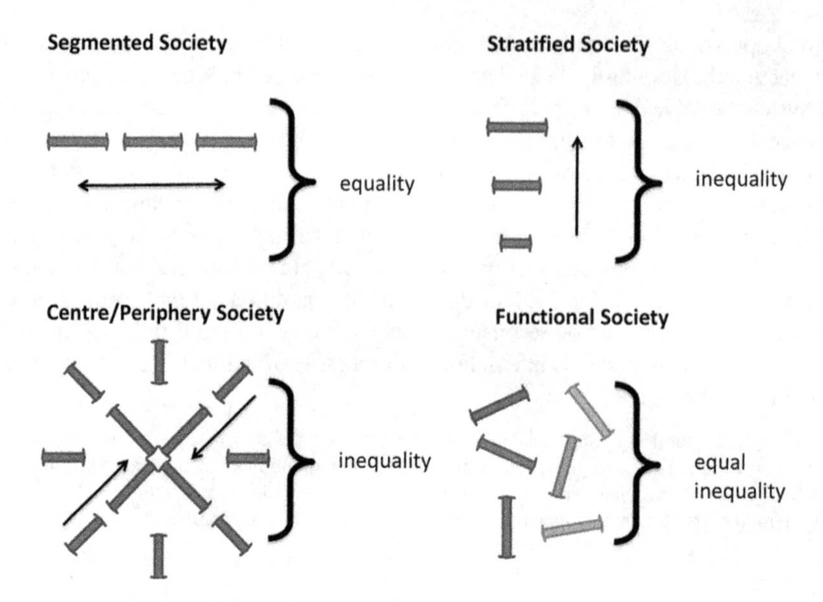

Fig. 4.4 Forms of social differentiation

of society. The reduction of complexity performed by society, allows these systems to build up more specific forms of communication[3] (Baraldi et al., 1996, pp. 154–155).

Luhmann identifies four main types of social differentiation throughout history (Luhmann, 2013, p. 12). In particular: (A) segmentary differentiation (equality between the partial systems); (B) centre/periphery differentiation (inequality between the partial systems, based on the proximity or distance from a centre); (C) stratified differentiation (inequality of the partial systems based on their position in a rank); (D) functional differentiation (equal inequality of the partial systems) (Fig. 4.4).

The different types of social differentiation are not mutually exclusive; on the contrary, they often co-exist and compete with each other. Yet, it is possible to identify "a dominant differentiation form in every societal system..", "the most important societal structure, which, if it can impose itself, determines the evolutionary possibilities of the system and influences the formation of norms, further differentiation, self-description of the system and so forth" (Luhmann, 2013, p. 11). The description of society that emerges from the theory of social systems is not based "on a unifying principle, a transcendental reference or a finalist purpose; society is described not on the basis of an underlying unity but on the basis of underlying difference" (Moeller, 2013, p. 40).

As a result, the concept of *modern society* proposed by Luhmann derives from its form of differentiation. "We understand modern society as a functionally differentiated society" (Luhmann, 2013, p. 87). This type of differentiation became dominant

[3] The translation from Spanish is mine.

between the sixteenth and eighteenth century and replaced stratified differentiation. Whereas, up until then, the main organizing principle of society had been the exclusive membership of a social strata (nobility, clergy, commoners) (Luhmann, 2013, p. 27):

> …in a functionally differentiated society, the partial systems are unequal because they have their own specific function. All partial systems are different and are defined on the basis of the function they develop within society. The main functional systems are: the political system, the economic system, the scientific system, the educational system, the law system, families, the religion system, the healthcare system, the art system (Baraldi et al., 1996, pp. 58–63).

All functional subsystems have evolved in their own particular way developing "their own set of symbolic codes, leading values, operational programs and regulative means" (Cvajner & Sciortino, 2010b, p. 392). In relation to individuals, function systems are in principle all-inclusive. This means that they include or exclude individuals only on the basis of their particular functional code, since they are indifferent to all other possible characteristics. For instance, the economic system only distinguishes between a convenient or inconvenient economic transaction, and does not take into consideration whether the participants are Algerian or Bolivian, lawyers or butchers, aristocrats or clergymen; the scientific system considers a communication only on the basis of its scientific value, and it is not concerned about whether the proponent comes from Ghana or Chile, is rich or poor, lawfully residing or not, etc. The same logic applies to every function system. In modern society the "chances to become included in different social realms – the economy, law, politics, education, health and the family – are no longer based on descent, or belonging to a social strata, or to an ethnic or religious group" (Bommes, 2012d, p. 37), and in this sense, there is no unitary principle of inclusion or exclusion.

> If society switches from stratification to functional differentiation, it also has to do without the demographic correlates of its internal differentiation pattern. It can then no longer distribute the people who contribute to communication among its subsystems as it had been able to do under stratification schema or centre-periphery differentiation. People cannot be attributed to functional systems in such a way that each belongs to only one system – the law, the economy, politics, the education system. Consequently, it can no longer be claimed that society consists of people; for people are clearly to be accommodated in no subsystem of society, and hence nowhere in society (Luhmann, 2013, p. 87).

Modern society is therefore "a complex multiplicity of a wide variety of system-environment realties without a centre, an essential core or a hierarchy" (Moeller, 2013, p. 24). Within its realm, no subsystem can claim to present the whole picture, but everyone produces an interpretation of the whole society.

> Each functional system can fulfil only its own function. In an emergency, no system can step in for another even in a supportive or supplementary capacity. In the event of a government crisis, science cannot help out with truths. Politics has no capacity of its own to devise the success of the economy, however much it might depend on this success politically and however much it acts as if it could. The economy can involve science in conditioning money payments, but however much money it deploys, it cannot produce truths. With financial prospects you can entice, you can irritate, but you can prove nothing (Luhmann, 2013, p. 99).

4.2.4 *Modern Society as World Society*

"With functional differentiation as its structural characteristic society is no longer primarily divided by regional borders, society is now a world-society" (Moeller, 2013, p. 52). As pointed out by Luhmann, with the rise of functional differentiation, there:

> "vanished those premises that enabled earlier social formations to include in the systems boundaries both the relation between systems and environment and those among different systems. Today we cannot expect that the differences between systems and environment and relation among different systems converge in one single system (the political) boundary" (Luhmann, 1982a, p. 239) In this sense, "As a general rule we can say that territorial borders no longer limit entire societies, but only political systems (with all that belongs to them: in particular jurisdiction). Territorial borders have the task of differentiating the world society into segmentary political functional systems: that is in equal states" (Luhmann, 1982a, p. 240).

To make this point clear, it is important to underline a crucial point in Luhmann's theorization. While all the other function systems have a global reach, the political system, in order to better fulfil its function, namely, "the capacity to produce collectively binding decisions" (Luhmann, 2009, p. 143), is segmentally divided into territorial states. "The political authority of the nation-state ends at its borders" (Moeller, 2013, p. 53).

> "Basing itself on this form of functional differentiation, modern society has become a completely new type of system, building up an unprecedented degree of complexity. The boundaries of its subsystems can no longer be integrated by common territorial frontiers. Only the political subsystem continues to use such frontiers, because segmentation into "states" appears to be the best way to organize its own function. But other subsystems like science or the economy spread over the globe. It is therefore impossible to limit society as a whole by territorial boundaries, and consequently it no longer makes sense to speak of "modern societies in the plural..." (Luhmann, 1982b, p. 178).

This peculiar characteristic of the political system helps to explain the development of the nation-state semantics and its interpretation of the world as if it was divided into national societies (Luhmann, 2009, p. 217). Yet, the boundaries of the other subsystems, for instance, the economy, mass media, science, etc. cannot be integrated into territorial frontiers, as they transcend geography and politics.

From the perspective of social systems theory, the concept of globalization refers to the world-society and to the global reach of its function systems. However, in Luhmann's opinion, the idea of a unique society must not be confused with the idea of homogeneity. "Global society is a complex multiplicity of subsystems which are not integrated in an overarching global unity" (Moeller, 2013, p. 54). In this sense, while the effects of functional differentiation spread all over the globe, these effects "combine, reinforce and inhibit one another due to conditions that occur only regionally, and consequently generate widely differing patterns" (Luhmann, 2013, p. 128).

> These special local conditions may be structural couplings that promote a surge in modernization in the direction of functional differentiation. More typically, however, the autopoietic autonomy of functional systems is blocked or limited to sectors of its operational possibilities.

> It would at any rate be quite unrealistic to see the primacy of functional differentiation as self-realization secured by the principle. [...] From this point of view, functional differentiation is not the condition of the possibility of system operations but rather the possibility of their conditioning. This also gives rise to a systemic dynamics that leads to extremely dissimilar developments within world society. The regions therefore find themselves far from any macrosocietal equilibrium, and precisely in this context are presented whit opportunities by a destiny of their own, which cannot be seen as a sort of micro-edition of the functional differentiation form principle (Luhmann, 2013, p. 131).

4.2.5 The State Beyond Modern State Semantics

Within society, it is possible to identify another type of systems: the *organizations*. These include, for instance, schools, associations, companies, political parties, etc. These systems are closely related to function-systems and share with them crucial characteristics such as autopoiesis and operational closure. However, they display an essential difference. While function systems are all-inclusive, meaning that nobody can be excluded from participating in them, and thus they have a global reach, organizations establish a clear member/non-member distinction, and so are smaller and localized (Luhmann, 2009, p. 243). An example can be clarifying: whereas everybody can be educated, only a registered student can go to a particular school; while everybody can perform an economic transaction, only accredited brokers can buy and sell on the stock market.

The advantage of organizations, and their usefulness, relies on their ability to coordinate more effectively the internal processes of a system in order to accomplish a specific function. Membership is an evolutionary development that helps the attainment of this objective. By establishing stable, regulated relations between members and the system, it allows complexity reduction and higher degrees of rationality. Every organization adopts its own codes, rules and programmes, establishes participation requirements and builds up internal structures to take binding decisions. At the same time, as a counterpart, it provides certain services and advantages that are reserved to its members.

States are a specific type of organization, closely related to the functioning of the political system. Yet, the two cannot be confused (Luhmann, 2009, p. 254). The political system, as seen before, is a function system that, in order to fulfil its function, is internally differentiated into territorial segments. This strategy is a pragmatic, evolutionary solution to the problem of extremely diversified regional conditions. The great variance of cultures, populations, economic possibilities, and development stages, in the different parts of the world, would make it impossible to provide collectively-binding decisions from a unique political centre.

> The seek for democratic consensus and the use of the minority/majority scheme, characteristic of the political decision process, could not be optimized from the heights of a global political system. In that case it would make no sense to participate to democracy, since the differences could not be properly represented. If votes would be quantitatively distributed, Hollanders would always be outnumbered by the Chinese, and the Portuguese by the Indian (Luhmann, 2009, p. 239).

Yet, given the complexity of the political function, the segmental division into territorial portions is not enough and a further step is required. The possibility to effectively communicate decisions that bind collectively has historically evolved into two main strategies: on the one hand, the capacity to force obedience when voluntary collaboration is excluded (the monopoly of the legitimate violence); on the other, the stimulation of voluntary collaboration through a *quid pro quo* logic (security, rights, welfare). Both strategies are obtainable only through organization.

A further important distinction in order to comprehend the functioning of the political system and the state is, in Luhmann's opinion, the one among politics, administration and public (Luhmann, 2009, p. 263). This perspective "allows analysing the power relations and correcting the official representation that understands power as purely hierarchical" (Luhmann, 2009, p. 264). Also in this case, the relationship among the three should not be interpreted through an input/output model but through a circular irritation/resonance model. The concept of "operation power circle" (Luhmann, 2009, p. 265) does not allow one to identify an initial moment or a dominating actor. Politicians take decisions, which are implemented by the administration, which are judged by the public, which elects politicians, etc. This chain creates a complex interdependence among the three, in which each actor needs to fulfil its function but cannot forget its interdependencies.

The state, then, is the organization that allows the political system to factually implement and "organize" a number of mechanisms to provide society with collectively-binding decisions. Citizenship, in turn, is the specific form of membership of this organization. Since it is the biggest and most complex social organization, the state is not a monolithic unity (Boswell, 2007), but is internally differentiated into a myriad of smaller structures and institutions. The relationship among these structures follows the irritation/resonance model, which explains the impossibility to locate entirely coherent, all-embracing, top-down decisions.

While closely entangled, the distinction between the concept of political system and that of state is fundamental for a number of reasons. (A) Whereas the state is a specific, developmental, historical solution to the requirements of the political system to fulfil its function, the relationship between the two is neither exclusive, nor fixed and unalterable. The capacity of the state to produce collectively-binding decisions can be disputed and, in some cases, a new organization can emerge as an alternative and take up the political function. (B) The state is an exclusive organization, and the benefits associated with membership usually apply only to its citizens; the political system is an all-inclusive function system and its communications apply to everyone within its territory. For this reason, for instance, although everybody can be arrested, only citizens can vote. (C) The citizen/non-citizen distinction is an organizational strategy of the state that helps the functioning of the political system. This strategy, however, is neither able to completely monopolize the political communications nor is it able to control the other systems within society. Regarding the first aspect, to give an example, if the state is not able to impose collectively-binding decisions on a group of non-citizens, or the application of certain rules, it may be forced to include those non-citizens or to change those rules. With regard to the second aspect, the state cannot completely limit the participation of non-citizens in other communications, such as, economic, scientific, artistic, etc. ones. (D) It is important to remember, that the state is not the only organization that is part of the political system and that helps the fulfilment of its function. Other organizations, such as, political parties, associations, syndicates, etc. may play an important role. A particularly interesting case within this group is that of international organizations. The European Union, the UN, the WTO, etc., for instance, are organizations that, usually with the agreement of states, have been acquiring powers in a number of sectors. The increasing importance and capacity of these organizations to produce collectively-binding decisions can be interpreted in relation to the globalizing effect on the political systems of functional differentiation (Moeller, 2013, p. 53).

Whereas for social systems theory the distinction between the state as an organization and the political system as a function system is structural, the semantics of the modern-state did not recognize this fact. In Luhmann's opinion, globalization evidences this point and helps to reveal:

> the secret premise of modern state thought: that of being the biggest and most efficient social organization and, together, the self-description formula of the political system. With the semantics of the state, a step was taken to put politics in the position to refer not only to the city or to the domestic context. The state recovers the expectation, included in the concept of civil society and *res publica*… […], to realize the unity of social order vis-à-vis the multiplicity of individual interests. When Carl Schmitt speaks of the end of statehood, […] he refers to the impossibility to maintain such pretension (Luhmann, 2009, p. 234).

These theoretical elements on the concept of the state have a number of implications. (A) The idea of the state as an autopoietic, operationally closed organization implies the possibility of immense differences in the particular strategies, characteristics and capacities that each one develops within complex and diversified regional system-environment configurations. (B) They imply that, in order to be able to produce collectively-binding decisions, every state has developed a particular mix of strategies that combine both deterring/threatening measures and encouraging/supportive ones. (C) The idea of the state as a dominating, leading actor within society is abandoned. As happens for all the other components of society, also the state relates to its environment through irritation/resonance relations. (D) While the irritations coming from the social environment, e.g. economic interests, humanitarian claims, mass-media pressures, public opinion, certainly resonate with its structures, the state operates and modulates its actions only in relation to its own functional imperatives (see, Boswell, 2007). (E) Since they are the biggest and most complex social organizations, states are internally differentiated into a myriad of smaller structures and institutions. The relationship among the internal structures follows the irritation/resonance model.

4.3 Irregular Migration as a Structural Phenomenon of World Society

4.3.1 Migration in World Society

In the conception of modern society offered by social systems theory, international migration, intended as the movement of people (migration) across state borders (international), appears as an inevitable, expected, structural phenomenon (Bommes, 2012c). Two elements concur to explain this fact. On the one hand, the rise of functional differentiation as the main type of social differentiation has determined the globalization of societal communications and the development of a unified world society. This has implied an increasing pressure on individuals to follow the inclusion opportunities offered by the different social systems (economy, education, family, science, religion, etc.) wherever they emerge. On the other hand, the particular form of differentiation adopted by the political system, i.e. the segmentation into territorial clusters, and the rise of states as the main form of political organization, have determined the enclosure of such societal opportunities within the sealed borders that divide each territory (Bommes & Sciortino, 2011a, p. 214). This has determined that, in order to access such opportunities, individuals need to cross political borders.

The particular configuration of modern society shows, then, a structural contradiction: while function systems are all-inclusive and foster human mobility across the world, the characteristics of the political system, namely states' territorial borders and exclusive membership, limit such mobility. In this sense, while international migration appears as an inevitable feature of world society, its existence is, nevertheless, problematic. As pointed out by Bommes:

> …migration is, on the one hand, probable as an attempt to take advantage of opportunities for inclusion. In terms of the economy, the law, education or health, and of modern organizations, migration is something individuals can be expected to do to adjust to the forms of inclusion they offer to them. Migration is therefore part of the normal, i.e. socially expected mobility in modern society, which has historically been implemented, for example, with the institutionalization of labour markets. The case of internal migrations within states' territories makes this clear. They are part of normal events that hardly mobilize social attention. Migration is, on the other hand, manifestly treated as improbable and as a problem, particularly in those countries with fully developed nation states and welfare states, when migration crossing state boundaries is involved (Bommes, 2012c, p. 27).

This scenario calls into question the specific characteristics of the political system and, in particular, of the organization that has monopolized its function, the state. This becomes evident, as suggested by Bommes, when internal migrations are considered. Also in this case, people decide to physically move in order to take advantage of better social opportunities, yet, since no political border is crossed, the phenomenon is unproblematic.

The state, like all the other organizations, uses a member/non-member distinction as a crucial strategy in order to fulfil its own function. In this case, membership helps the production of collectively-binding decisions in two main ways. On the one hand, it allows the state to register its members and enrol them in institutions, such as, the police or the army that factually permit the monopolization of the legitimate means of violence (Torpey, 1998). On the other hand, it makes the development of a mutually beneficial relationship based on the exchange between loyalty and service possible (Bommes & Geddes, 2000a). The state offers a number of different provisions (security, rights, assistance, etc.) and in turn receives individuals' fidelity and obedience. If this is the basic idea, a fundamental question needs to be answered: how are members selected? On what basis?

The particular type of political membership developed by the modern state, i.e. *national citizenship*, emerged as an evolutionary solution that was able to link in a seemingly natural, immediate and permanent way a single population (the nation), with a specific territory and a political sovereign (the state) (Bommes, 2012b; Halfmann, 2000). The construction of this link and its stabilization, anything but natural, required an immense effort by every state and was the cause behind many of the wars and conflicts that characterized modernity. This effort involved historians and politicians, soldiers and teachers, artists and businessmen, who help to develop a national sentiment among otherwise fragmented and differentiated populations (Benedict Anderson, 2006; Hobsbawm, 2012; Smith, 1986). Notwithstanding the difficulties, this conception of membership was able to develop, in a relatively short time, into a "particular universalism that envisages the inclusion of every individual into one, but only one state" (Bommes, 2012c, p. 27).

The flipside of this process was the creation of the figure of "the foreigner" as the natural counterpart of "the national". While the latter had the right to access the services offered by the state and to freely circulate in and out of its territory, the former was in principle excluded from every benefit and banned from entering the national borders without a valid permit. Thanks to these configurations, as suggested by Bommes and Geddes, states evolved into "thresholds of inequalities", since the communicational possibilities available within their borders became accessible, at least ideally, only to their citizens (Bommes & Geddes, 2000a). This tendency became more and more marked as states evolved into welfare states and the services and opportunities offered to their members constantly increased. To be a citizen of a rich state and not of a poor one, allowed incomparable access opportunities to function systems, such as, the economy, law, science, education, health, etc.

Yet, the all-inclusive character of functionally-differentiated social systems severely questioned the idea of immobile, confined populations, which was alleged within the nationalist conception. The "sedentary bias" (see Chap. 3) proved to be unrealistic and the figure of "the migrant" emerged in the very same moment in which territorial borders were drawn. Individuals, along with the development of world society, were increasingly stimulated to follow inclusion opportunities beyond the regulations and borders established by states. In this sense, migration can be interpreted as an effort made by individuals to achieve social inclusion, as a way to achieve social mobility through spatial mobility (Bommes & Sciortino, 2011b).

4.3.2 States and Migrants

Against this backdrop, the structural contradiction between migrants and states becomes evident. On the one hand, migrants try to "achieve inclusion and participation in the various social systems – and with them, access to the relevant social and economic resources – by means of geographical and border-crossing mobility" (Bommes & Sciortino, 2011b, p. 214). On the other hand, states try to reaffirm the basic mechanism of their functioning, i.e. the distinction between members and non-members which allows the loyalty/service exchange.

This contradiction, however, should not be interpreted in absolute, unconditional terms. While it is true that membership is the core feature of the state as an organization and that foregoing this could undermine its very existence, it should be borne in mind that the main function of the political system is not the distinction between members and non-members, but the production of collectively-binding decisions. In relation to this function, the membership strategy is certainly useful, but it is not the only one. In particular, if it is true that society is functionally differentiated, and that individuals seek inclusion in the different systems, the political system, in order to fulfil its own function, cannot impede the functioning of the other systems. If that were the case and other communications of other systems became obstructed, the possibility to produce collectively-binding decisions could be seriously undermined. Individuals would have strong incentives not to follow the decisions of a system that precludes all other communications. For this reason, states are pushed to develop ecological equilibriums with other systems through irritation/resonance relations (Sciortino, 2000). The same, of course, is valid for every system: each has to fulfil its own function but, in so doing, it observes and resonates in the relations with the others.

With regard to migration, while states use and defend the member/non-member distinction, and thus enact policies to control and limit the arrival and residence of foreigners or their access to the services, they must, at the same time, take into consideration the functioning of other systems, for instance, the economy, the law, the family, etc. If an economic sector requires unqualified workers who are not available in the internal market, for example, the political system could decide to amend its principles and admit migrants.

As becomes apparent, the relation between migrants and states is much more complex than the idea of a forthright contraposition might suggest. The perspective offered by social systems theory suggests that this relation embodies in a variety of national settings (Bommes & Geddes, 2000a). Each state, on the basis of its own particular political characteristics, organizational infrastructure, and public opinion, and in relation to both its intra-social (the other function systems) and extra-social environment (the effective migration process) develops a specific, historically-influenced approach to migration. The wide variety of policies analysed in Chap. 2, that range from external and internal controls to migrant labelling and categorization, from legalizations to expulsions, can be understood within this framework. States can be viewed as "political filters" (Bommes & Geddes, 2000b, p. 2) which mediate not only migrants' efforts to take advantage of their chances for social participation, but also other system demands for migrants.

The different approaches taken by states in relation to migration can also be related to their greater or lesser desire (and capacity) to penetrate their society. As pointed out by Bommes: "nations-states cannot renounce their right to control access to and residence to their territories. This right is implemented very differently: from states' wide-ranging, deep social penetration to lighter and more limited approaches" (Bommes, 2012a, p. 166). In relation to this issue, the development of the modern welfare state is particularly relevant (Bommes, 2012d; Bommes & Geddes, 2000a; Halfmann, 2000; Sciortino, 2004b). The increased services offered to their citizens, in connection to the evolution of the conception of rights (from political to civil, to social), implied a continuous expansion of the state's influence within society. Although this development took diverse paths in the different areas of the world, it has been possible to identify certain patterns and to produce welfare-state typologies (Esping-Andersen, 1990, 1996; Ferrera, 1996; Ferrera, Hemerijck, & Rhodes, 2000; Hemerijck, 2012). In all cases, the involvement of the state in more and more social sectors (education, healthcare, pensions, unemployment support, etc.) and the provision of increasingly-sophisticated and costly services required the extension of the member/non-member logic to each of the new domains of state intervention. It is precisely this latter aspect that further complicates the relation between states and migrants. As pointed out by Sciortino, this occurred in much wider terms than those suggested by the welfare magnet thesis (the welfare state attracts migrants) or by the welfare dependency debate (do migrants contribute to or exploit welfare?) (see Sciortino, 2004b). Recalling Esping-Andersen's approach, he points out that "the welfare structures must be considered as embedded in a matrix of structural relationships among households, the state and the economy. It is precisely within this framework that the relationships between welfare structures and migratory processes may be investigated in full" (Sciortino, 2004b, p. 115). In particular, the extent and specific ways in which state intervention alters the functioning of the other social systems, and the modes in which the political distinction members/non-members penetrates other social realms, can deeply influence the migratory phenomenon. Depending on the case and on the sector, this influence can have the effect of fostering or discouraging migration, of favouring certain types instead of others, of creating better or worse conditions for migrants' inclusion. The differential analysis of welfare regimes is, then, a crucial requirement in order to comprehend not only the interaction between migration and the state, but, more in general, between migration and society.

To make the picture even more complex, it is important to consider three additional issues. (A) States are internally differentiated and each section of their enormous apparatus can develop a certain intra-organizational vision of migration. This implies that monolithically-coherent, one-directional decisions tend to be the exception while diversified, conflictive and multi-levelled ones are the rule. (B) Not only is states' internal view on migration fragmented and conflictive, but also that of the environment. The various social systems may have very different interests concerning migration; therefore, the state is usually confronted by a large number of often un-reconcilable demands (Boswell, 2007). (C) While the modern state semantics offers the idea of the political system as a regulator of society, as a predominant actor capable of controlling and steering every social process, this is only a self-description. In relation to migration, this means that no state, not even the most developed and determined one, is able to perfectly manage population movements (Cvajner & Sciortino, 2010b, p. 394).

4.3.3 Irregular Migration as a Structural Phenomenon of World Society

Irregular migration is probably the social phenomenon that best highlights world society's structural contradiction between the global, all-inclusive, functional characteristics of all the other social systems and the territorially-bounded, exclusive, segmented characteristics of the political system (Bommes, 2012a, 2012c; Bommes & Sciortino, 2011a; Cvajner & Sciortino, 2010a, 2010b). From the perspective of social systems theory, the emergence of irregular migration must be understood as a logical outcome, embedded in the structure of contemporary society.

> At the root of the contemporary migration system is a structural mismatch between the huge demand for entry to the most developed regions and the comparatively small supply of opportunities to enter these areas legally. It can consequently be described as a social system – a structured nexus of interdependencies – where there is an embedded tension within the cultural and social goals prescribed by an increasingly shared global culture and the means available to pursue these goals (Bommes & Sciortino, 2011b, p. 215).

Within this context, migrants are faced with two contradictory communications. On the one hand, function systems, such as, the economy, education, family, etc., which do not recognize the national/foreigner distinction, offer opportunities that attract them. On the other hand, the political system and its main organization, the state, demand membership to allow entry, and therefore discourage their movement. Confronted by this double message, "come/do not come", the migrants' decision is the result of a complex evaluation of pros and cons. The political limitations imposed by states, although important, are only one of the issues at stake. If the opportunities are great enough and there are no regular channels available, the option to migrate irregularly becomes a valid and sometimes unavoidable alternative.

The birth and development of irregular migration systems is contingent upon the existence of a structural mismatch between the social and the political conditions for migration. As pointed out by Sciortino, such mismatch involves both sending and receiving contexts, and it has both an external and an internal dimension. "Externally, there must be a mismatch between the demand for entry, and the supply of entry slots by the political systems" in the receiving context. In the sending one, "there must be a mismatch between widespread social expectations (usually called "push factors") and the state capacity to satisfy them or repress them". In the receiving context, "there must be a mismatch between the internal pre-conditions for migration (usually called "pull factors") and their interpretation within the political system. Irregular migrations are in fact an adaptive answer to these unbalances" (Sciortino, 2004a, p. 23).

The existence of such mismatches and the emergence of irregular migration have been interpreted by Bommes and Sciortino in connection to Merton's concept of "structural anomie" (Merton, 1968). As they underline:

> "...an emphasis on social structures as regulators of individual behaviour does not imply that social structures are not also involved in determining the circumstances in which the violation of established social norms is 'normal' – that is, predictable in terms of their contradictory device". "...both conformity and various types of deviance should be seen as adaptive strategies to deal with the structural mismatch between prescribed goals and institutionalized means in a society prizing economic success and social mobility as attainable by all its members. If we apply Merton's framework to the current world migratory situation, we can conclude that irregular migration is actually a specific form of innovative behaviour. It represents a creative solution to the structural mismatches inherent in modern society – i.e. the demand for labour and the available supply of workers or the demand for social mobility and the supply of opportunities for advancement. It is a strategy that implies breaking away from the use of institutionally prescribed (but obstructed) means in order to keep a communal faith and commitment to the culturally and socially prescribed goals increasingly shared in sending and receiving areas" (Bommes & Sciortino, 2011b, p. 216).

This conception helps to understand a particular feature of irregular migration, otherwise interpretable only in paradoxical or conspiracy terms. From a logical perspective, it could be said that it is the state itself that, by establishing entry criteria and distinguishing between regular and irregular migrants, creates the problem that it later tries to solve. Yet, here it is important to bear in mind that when a system irritates its environment on the basis of its own logic and seeking its own purposes, the environment resonates on its own terms, according to its own logic and in relation to its own purposes. In this sense, while the concept of "legal production of irregularity" may be factually true, its interpretation in terms of a state's intention or hidden strategy supposes a capacity to control its environment that is unrealistic.

This does not mean denying that state actions may create the conditions for irregularity to develop and evolve. The goal is to warn against simplistic, lineal, cause/effect conclusions. As pointed out by Bommes, for irregular migrants:

> ...opportunities to participate arise in labour markets, families and elsewhere, and gain greater permanency because there is a receptive context for them, one which is in part politically and legally constituted by the same welfare states which seek to control and prevent these migrations. This is not meant only in the trivial sense that everything which is illegal about illegal immigration is only illegal because there are corresponding laws which limit or prohibit residence or work, but more particularly in the sense that motives arise in labour markets, in private households, in housing markets or in welfare organizations themselves to disregard such limitations or to use them as boundary conditions for establishing employment relations and tenancies, for starting families, providing services or setting up aid organizations which would scarcely come about otherwise (Bommes, 2012a, p. 160).

The structural character of the irregular migration phenomenon does not imply that it occurs in a smooth, non-conflictive way, but quite the contrary. On the one hand, state efforts and capacity to control irregular migration have increased enormously in the last decades. States' knowledge of the phenomenon has constantly increased, allowing the adoption of more sophisticated strategies. Yet, these efforts, as the theory of social systems suggests, have never been able to fully regulate the other social processes. "States' claim of control over a territory is just a claim within various, but never with complete degrees of implementation. Strong mechanisms of control fail when the opportunities to be gained through migration are strong and the social pre-condition for migration amply fulfilled" (Sciortino, 2004a, p. 22). On the other hand, also migrants develop their strategies, increment their knowledge, and build up their infrastructures. This allows them to circumvent state controls, although at very high costs.

4.3.4 *Irregular Migration as a Differentiated Sociological Phenomenon*

To understand how an irregular migration phenomenon initiates and develops, which resources it mobilizes and what structures and interactions it establishes, it is necessary to consider the dynamic interplay not only between states and migrants, but also between these and all the other social systems. Each actor needs to be considered as internally differentiated, self-referential and, yet, deeply interrelated with its environment through irritation/resonance relations. The main consequence of this radically differential perspective is that the particular phenomenology of each "irregular migration reality" cannot be theoretically or legally deduced, but it must be empirically researched. In this sense, whereas in legal terms it may be possible to talk about irregular migration as a single category, from a sociological perspective, it is more accurate to talk about *irregularities*. In each context, the systemic interactions among states, migrants and the other social systems set the conditions for the emergence and evolution of differentiated irregular migration realities. This approach has a number of theoretical and methodological implications.

Irregular Migration as a Status

The irregular status, attached to migrants by the political system, does not describe their whole social position. "From the point of view of systems theory, individuals are not part of society and therefore also not integrated or 'incorporated' into society" (Bommes, 2012c, p. 25). The relationship between individuals and society based on the concept of differential functional inclusion makes the question about the opportunities of irregular migrants empirical. The questions, then, become: How are irregular migrants included in the different social systems? How does the exclusion from political membership affect other inclusions? As stressed by Bommes and Sciortino: "in modern society there is no full total identity, the status is only one piece of the puzzle that is composed by a variety of statuses variously significant in different contexts" (Bommes & Sciortino, 2011b, p. 219). This condition may imply that irregular status, usually interpreted only as excluding, can turn out to be a condition for inclusion. In certain contexts, for instance, the irregular status may favour the inclusion in the economic system. This evidences how the exclusion from state membership does not necessarily prevent irregular migrants from participating in the other social systems.

Irregular Migration and States

The relation between politics and irregular migration cannot be interpreted in linear, straightforward, oppositional terms (the state vs. irregular migrants). There are different reasons for why this is so. Firstly, states must be considered as internally diversified "organization complexes" (Bommes, 2012d) composed of a wide variety of institutions, agencies, departments, bureaucracies and levels of government. Moreover, the political functioning must be considered in terms of a "power cycle" in which politics, administration and the public reciprocally influence and legitimate each other. Therefore, as happens for most political issues, also for irregular migration, a single, coherent, stand is not available; each component develops a pragmatic approach in an attempt to fulfil its own particular duty and to remain legitimate. This may imply phenomena like the coexistence of policies that favour and disfavour irregular migration, the development of legal loopholes, policy inconsistency along the decision chain, etc. Secondly, it should be borne in mind that, while the member/non-member distinction is an important element of the functioning of states, their core function is the attainment of politically-binding decisions. In this sense, although the control of irregular migration is, in principle, of great importance, the fulfilment of the function is even more relevant. Accordingly, depending on the specific context, the historic moment, the effective capacity to implement policies and the demands coming from the other systems, states may decide to be flexible as regards the membership principle and choose pragmatic approaches that may include: turning a blind eye, the use of symbolic policies, mass-legalizations, etc. Thirdly, whereas states are powerful organizations and the political system plays an important role within social communications, neither of them is capable of dominating society and of completely controlling other system transactions. Adopting a differential perspective, a state's degree of social penetration and policy implementation capacity becomes an empirical question that has to be answered after analysing each case. Depending, for instance, on the different political traditions and regimes, the type of welfare, the administration's degrees of development and cultures, public positions and levels of concern, state policies may be very different, and likewise their impact on irregular migration. As pointed out by Bommes and Sciortino, the amount and types of transactions where legitimate residence is considered significant, and the capacity by states to effectively check it, can dramatically change the meaning of being irregular (Bommes & Sciortino, 2011b, p. 217).

Irregular Migration and Society

The systemic understanding of society not only excludes the possibility of political systems to dominate social transactions; it also excludes that of every other system. Accordingly, neither the economic system nor the legal one, neither the familial nor the educational one, just to mention some, can exert control over society and impose their logic. The reality of irregular migration can be interpreted as the result of the dynamic interplay among the different approaches, interests, and concerns of each system. In this sense, while each system produces its own interpretation, the phenomenon cannot be fully understood only on the basis of one of these.

Irregular Migrants

Even within a single country, the irregular migration phenomenon must be considered as dynamic and internally differentiated. Migrants' interactions with states and with the other systems produce a myriad of different migration trajectories (Sciortino, 2004a, p. 38). This can be related to a number of factors. Firstly, it may be linked to the enormous differences existing between different groups of migrants and between individuals within each group. The availability of human, social and economic capital can make a paramount difference, especially with regard to irregular migrants, since their effort is more complex and cannot count on the support of states. Secondly, the time factor plays a crucial role. The success of an irregular migration trajectory is related to the ability of migrants to analyse the environment and to develop strategies and counter-strategies to deal with problems. These strategies are necessary, for instance, to avoid controls, discover and take advantage of possible legal loopholes or to develop specific social structures. Since there is no instruction booklet available and the social environment continuously changes, irregular migrants need to rely on a learning-by-doing approach and on the development of a trusted network. In both cases, time makes a big difference. The concept of "migratory career" proposed by Cvajner and Sciortino, and derived from Luhmann's theory offers an adequate tool to analyse irregular migrants' trajectories. Intended as "a sequence of steps, marked by events defined as significant within the structure of actors' narratives and publicly recognized as such by various audiences" (Cvajner & Sciortino, 2010a), the notion makes it possible to follow the experience of individual irregular migrants and to identify possible common patterns within a similar migratory context.

4.4 Conclusion. A Systemic Analytical Framework for Irregular Migration

Irregular migration has usually been interpreted either through the lenses of states or through the lenses of migrants. This has generated two main perspectives on the phenomenon: the first understands it as a problem that may signal an erosion of states' prerogatives; the second understands it as a form of exploitation, which signals states' enduring capacity to seek their goals. Although they contrast each other, both perspectives are based on a similar, problematic, conception of society, social actors and social relations. This conception, based on the semantics of modern states, understands society as subsumed within the concept of the state. The latter is conceptualized as a predominant actor that is able to control (or lose control over) the former. Social actors are intended as monolithic, single-minded, and time-stable players. Finally, social relations are interpreted through an input/output model that, accordingly, presupposes the possibility to establish clear-cut, cause/effect interactions. Irregularity, from this standpoint, is understood as a rather undifferentiated phenomenon that, depending on the case, signals either a state effective strategy or failure.

Niklas Luhmann's social systems theory proposes a radical critique of the semantics of modern states. Society, in this conception, regains a central, all-embracing role. The political system and the state, although important, are considered as only two among the numerous systems and organizations constituting the complex galaxy of social relations. On the basis of this notion, irregular migration should be understood as a complex, differentiated, structural phenomenon of modern world society. The development of this phenomenon is related to the existing structural mismatch between the dominant form of social differentiation (functional) and the specific form of internal differentiation (segmentary) into territorial states of the political system. This creates a fundamental conflict between two logics: on the one hand, the all-inclusive logic of most social systems (economic, legal, educational, familial, etc.) that fosters human mobility across geographic space; on the other, the exclusive logic of states that insists on regulating human mobility on the basis of a membership principle. Against this backdrop, irregular migration emerges as an adaptive solution to the mismatch existing between the high demand for entry into certain states and the limited number of legal entry slots available.

If, in abstract and theoretical terms, irregular migration is explained as a structural feature of world society, the concrete, sociological manifestations embodied in the phenomenon within each context cannot be theoretically deduced. Instead, irregular migration realities must be empirically researched and understood as the result of a context-specific, dynamic, evolutionary interplay among: (A) functional social systems; (B) states; and (C) migrants. As suggested by the theory of social systems, each actor needs to be considered as autopoietic, self-referential and internally differentiated; social relations must be interpreted through an irritation/resonance model instead of an input/output model.

Irregular migration realities can be understood, then, as the result of a complex "equation of irregularity" (Arango, 1992, 2005; Arango & Finotelli, 2009, p. 16) that ponders the role of different actors involved and the many variables at stake. Table 4.1 presents a non-exhaustive analytical framework of the relevant actors and variables affecting the generation of irregular migration realities. In every context, the specific "weight" of every actor, the value of every "variable" and the particular relation among all these factors produce a different result. This transforms into a different ecological positioning of irregularity with regard to the rest of society and into a number of different irregular migration careers developed by migrants.

Table 4.1 Systemic analytical framework for irregular migration realities

Actors				Variables
STRUCTURAL CONTEXT	**Political system**	States	Politics	Type of political regime Type and levels of services (welfare regime) Political and migration culture Geographical accessibility and proximity of migration sources
			Administration	Extension and efficiency Administrative culture and tradition Internal differentiation and level of government
			Public	Ideologies Civic and migration culture Concern versus migrations
		Other political organizations	Political Parties Organizations Syndicates	Ideologies Civic and migration culture Concern versus migrations
		International organizations	EU, UN, IOM, UNHCR, ILO, etc.	International agreements, decision structures, provisions
	Economic system	Productive structure		Main economic sectors Labour market structure Underground economy
		Economic dynamics		General economic trends Sectorial economic trends
	Legal system	Internal		Legislation regarding migrations (entry, residency, naturalization, regularization, labour market, welfare services entitlements and access, territory control, etc.) Structure and functioning of legal systems
		External		International legislation Structure and functioning of international legal system
	Family system			Family structure and distribution Familial ties and supportive structure

(continued)

Table 4.1 (continued)

Actors			Variables
	Education system		Educational levels and accessibility in origin and destiny Role of the public institutions and existence of alternatives
	Health system		Health care levels and accessibility in origin and destiny Role of the public institutions and existence of alternatives.
	Mass media system		Culture transmission Transmission of opportunities and options Communication on migration (concerned, indifferent, positive)
	Religion system		Religious view of migration Religious support structures
	Migrants' social structures		Network structures and activities Illegal network structures and activities
MIGRANTS	**Migrants' capital**	Social	Networks (types, extension, functioning)
		Cultural	Languages, professions, communication abilities, etc.
		Economic	Money
	Numbers		Irregular migrant numbers
	Time		Migration length
	Type of migration		Permanent, circular

4.4.1 What Advantages?

As pointed out in the conclusions of Chap. 3, the theoretical understanding of irregular migration presents two main problems: the treatment of irregular migration as an undifferentiated phenomenon and the use of mono-causal explanations. The proposed explanations, moreover, appeared difficult to reconcile, and were even quite contradictory. These problems were connected to three crucial theoretical flaws common to the majority of the analysed theories, in particular: (A) the state-centric conception of society; (B) the limited conception of the different social actors; (C) the inadequate understanding of social interactions.

The theoretical elements gathered from Niklas Luhmann's theory of social systems, offer interesting, possible improvements to the three theoretical problems pointed out. In connection with these improvements, the resulting understanding of irregular migration appears more complex but, arguably, more consistent with reality.

A number of theoretical and methodological advantages can be suggested:

(a) Irregular migration is not understood as a state strategy, as a migrant's tactic, or as an economic advantage, but it is understood as a social product resulting from the complex and dynamic interaction between all social systems. Whereas all social actors create their own perspective of the phenomenon and display their own interests, approaches and concerns, the overall social significance of irregularity results cannot be deduced only from one of them.
(b) Irregular migration is understood in a radically differentiated way. Its concrete forms, structures, social relevance, evolution, externalities are determined by the context-specific configuration that the phenomenon adopts. From a systemic sociological perspective, therefore, it is not possible to understand irregularity as a single phenomenon, but rather as a multiplicity of irregular migration realities.
(c) The role of the state in relation to irregular migration is understood in a less deterministic way. Since states are not able to fully control and determine social transactions, they are neither omnipotent nor helpless. Each state, depending on a number of variables, is more or less able to enforce its decisions. The way in which these decisions resonate with the other social systems and with migrants is not in its hands and must be empirically researched.
(d) State policies are not considered as necessarily and coherently against irregular migration. This responds to two factors. A. The very complex forms of states' internal differentiation, which entail the possibility of phenomena like the coexistence of policies that favour and disfavour irregular migration, the development of legal loopholes, policy inconsistency along the decision chain, etc. B. States are organizations that use the member/non-member principle in order to better fulfil the political system's function. Yet, this function, namely the production of collectively-binding decisions, in certain cases or thanks to the interaction with other social systems, can be better fulfilled with a flexible understanding of the membership principle. Pragmatic solutions, that may include turning a blind eye, the use of symbolic policies, mass-legalizations, etc., can be understood in relation to the resonance relations of the state with the other social systems. The orthodox application of the membership principle could determine heavy externalities on the other systems, which may in turn have negative effects on the states' capacity to fulfil their own function.
(e) Irregular migration is understood as internally differentiated also within a national context. A number of factors, such as, migrants' origin, social, cultural and economic capital, migration duration, availability of migrant supportive (legal or illegal) structures, etc., may determine very different irregular migration careers. These can differ in terms of: (A) The amount and type of inclusions

within the different social systems (economy, education, health, family, religion, politics etc.); (B) The type of irregularity (for instance, permanent or circular); (C) The duration of the irregularity condition; (D) The social conditions.

(f) The irregular status is not understood as describing the whole social position of a migrant. The relationship between individuals and society is understood through the concept of differential functional inclusion. While irregularity describes the relation between migrants and the state as an organization, their inclusions in the other systems and the way in which that is affected by the political status is not politically determined.

A systemic theory of irregular migration allows one to understand the phenomenon as a radically differentiated, structural outcome of modern world society. Once the idea is disregarded that any actor or institution can control all social transactions, the whole focus changes. The query is no longer about actors' real intentions or covert plans, failure or success, domination or irrelevance; instead, it is about actors' decision-making processes and compromises, degrees of success or disappointment, and complex and dynamic interactions. While this hermeneutic approach would certainly offer less deterministic and clear-cut accounts of irregular migration, its multi-causal and differentiated explanations would certainly reach the aim of being more congruous with social reality.

This approach suggests the need to research irregular migration realities within each context. The possibility to discover common patterns and trends requires then an effort of comparative analysis. Only comparing the way in which irregular migration realities are conformed, develop and interact with their contexts, it will be possible to reach a deeper understanding of the phenomenon. Moreover, as suggested by Bommes, discovering the specific role and significance of irregularity within a context, it is also a way to better understand that context (Bommes, 2012a).

Bibliography

Anderson, B. (2006). *Imagined communities: Reflections on the origin and spread of nationalism*. London: Verso Books.

Arango, J. (1992). Los dilemas de las políticas de inmigración en Europa. *Cuenta y Razón, 73*, 46–54.

Arango, J. (2005). La inmigración en España: demografía, sociología y economía. In R. del Águila (Ed.), *Inmigración: Un desafío para España* (pp. 247–273). Madrid, Spain: Editorial Pablo Iglesias.

Arango, J., & Finotelli, C. (2009). *Past and future challenges of a Southern European migration regime: The Spanish case*.

Baraldi, C., Corsi, G., & Esposito, E. (1996). *Glosario sobre la teoria social de Niklas Luhmann*. Mexico, Mexico: Universidad Iberoamericana-ITESO – Anthropos Editorial.

Bommes, M. (2012a). Illegal migration in modern society: Consequences and problems of national European migration policies. In C. Boswell & G. D'Amato (Eds.), *Immigration and social systems. Collected essays of Michael Bommes* (pp. 157–176). Amsterdam: Amsterdam University Press.

Bommes, M. (2012b). *Immigration and social systems: Collected essays of Michael Bommes*. Amsterdam: Amsterdam University Press.
Bommes, M. (2012c). Migration in modern society. In C. Boswell & G. D'Amato (Eds.), *Immigration and social systems. Collected essays of Michael Bommes* (pp. 19–36). Amsterdam: Amsterdam University Press.
Bommes, M. (2012d). National welfare state, biography and migration: Labour migrants, ethnic Germans and the re-ascription of welfare state membership. In C. Boswell & G. D'Amato (Eds.), *Immigration and social systems. Collected essays of Michael Bommes* (pp. 37–58). Amsterdam: Amsterdam University Press.
Bommes, M., & Geddes, A. (Eds.). (2000a). *Immigration and welfare. Challenging the borders of the welfare state*. London/New York: Routledge.
Bommes, M., & Geddes, A. (2000b). Introduction: Immigration and the welfare state. In M. Bommes & A. Geddes (Eds.), *Immigration and welfare. Challenging the borders of the welfare state* (pp. 1–12). London/New York: Routledge.
Bommes, M., & Sciortino, G. (Eds.). (2011a). *Foggy social structures: Irregular migration, European labour markets and the welfare state*. Amsterdam: Amsterdam University Press.
Bommes, M., & Sciortino, G. (2011b). In lieu of a conclusion: Steps towards a conceptual framework for the study of irregular migration. In M. Bommes & G. Sciortino (Eds.), *Foggy social structures: Irregular migration, European labour markets and the welfare state* (pp. 213–228). Amsterdam: Amsterdam University Press.
Boswell, C. (2007). Theorizing migration policy: Is there a third way? *International Migration Review, 41*(1), 75–100.
Cvajner, M., & Sciortino, G. (2010a). A tale of networks and policies: Prolegomena to an analysis of irregular migration careers and their developmental paths. *Population, Space and Place, 16*(3), 213–225.
Cvajner, M., & Sciortino, G. (2010b). Theorizing irregular migration: The control of spatial mobility in differentiated societies. *European Journal of Social Theory, 13*(3), 389–404.
Esping-Andersen, G. (1990). *The three worlds of welfare capitalism*. Princeton NJ: Princeton University Press.
Esping-Andersen, G. (1996). After the golden age? Welfare state dilemmas in a global economy. In G. Esping-Andersen (Ed.), *Welfare states in transition: National adaptations in global economies* (pp. 1–31). London: Sage.
Ferrera, M. (1996). The "Southern model" of welfare in social Europe. *Journal of European Social Policy, 6*(1), 17–37.
Ferrera, M., Hemerijck, A., & Rhodes, M. (2000). *The future of social Europe: Recasting work and welfare in the new economy*. Oeiras, Portugal: Celta Editora Oeiras.
Halfmann, J. (2000). Welfare and territory. In M. Bommes & A. Geddes (Eds.), *Immigration and welfare. Challenging the borders of the welfare state*. London/New York: Routledge.
Hemerijck, A. (2012). *Changing welfare states*. Oxford, UK: Oxford University Press.
Hobsbawm, E. J. (2012). *Nations and nationalism since 1780: Programme, myth, reality*. Cambridge, UK: Cambridge University Press.
Luhmann, N. (1982a). Territorial borders as system boundaries. In R. Strassoldo & G. Delli Zotti (Eds.), *Cooperation and conflict in border areas* (pp. 235–244). Milano: Franco Angeli Editori.
Luhmann, N. (1982b). *The differentiation of society*. New York: Columbia University Press.
Luhmann, N. (1990). *Political theory in the welfare state*. Berlin, Germany/New York: Walter de Gruyter.
Luhmann, N. (2002). *Einführung in die Systemtheorie*. Heidelberg, Germany: Carl-Auer-Systeme.
Luhmann, N. (2006). System as difference. *Organization, 13*(1), 37–57.
Luhmann, N. (2007). *La sociedad de la sociedad*. México, México: Universidad Iberoamericana.
Luhmann, N. (2009). In J. T. Nafarrate (Ed.), *Niklas Luhmann: la política como sistema*. México. D.F., México: Universidad Iberoamericana.
Luhmann, N. (2012). *Theory of society – Volume 1* (Vol. 1). Palo Alto, CA: Stanford University Press.

Luhmann, N. (2013). *Theory of society – Volume 2* (Vol. 2). Palo Alto, CA: Stanford University Press.

Merton, R. K. (1968). *Social theory and social structure*. New York: Simon and Schuster.

Moeller, H.-G. (2013). *Luhmann explained: From souls to systems*. Chicago: Open Court.

Schinkel, W. (2010). The virtualization of citizenship. *Critical Sociology, 36*(2), 265–283. https://doi.org/10.1177/0896920509357506

Sciortino, G. (2000). Toward a political sociology of entry policies: Conceptual problems and theoretical proposals. *Journal of Ethnic and Migration Studies, 26*(2), 213–228.

Sciortino, G. (2004a). Between phantoms and necessary evils. Some critical points in the study of irregular migrations to Western Europe. *IMIS-Beiträge, 24*, 17–43.

Sciortino, G. (2004b). Immigration in a Mediterranean welfare state: The Italian experience in comparative perspective. *Journal of Comparative Policy Analysis: Research and Practice, 6*(2), 111–129.

Smith, A. D. (1986). *The ethnic origins of nations*. Oxford, UK: Basil Blackwell.

Torpey, J. (1998). Coming and going: On the state monopolization of the legitimate "means of movement". *Sociological Theory, 16*(3), 239–259.

Part II
Empirical Study

Chapter 5
Methodological Note

The second part of the book will present the results of an empirical study of the experience of Ecuadorian irregular migrants in the cities of Amsterdam and Madrid. While the general overview of the methodological conceptualization that is the backbone of the whole research work has already been outlined in the introduction, in this brief chapter the focus will be centred on the specific methodological issues concerning the empirical study.

5.1 Research Design and Research Questions

In the first part of this study, a number of critical aspects that have affected the theoretical understating of irregular migration came to light. Here to two of these will be further discussed: a tendency towards non-differentiation and then a tendency to the use of mono-causal explanation.

Regarding the first issue, there has been a tendency to treat irregular migration as if it were a single, undifferentiated phenomenon across different countries and historical phases. In Chap. 3, a wide number of possible explanations of this problem were discussed, especially from a conceptual/theoretical perspective. The theoretical framework derived from social systems theory in Chap. 4 attempted to conceptualize irregular migration as a differentiated phenomenon that develops specific characteristics and different forms, depending on the social context in which it develops. Yet, the problem of non-differentiation has not only derived from the theoretical (mis-)understanding of irregular migration. The problem has also been a consequence of the methodology used to empirically study the phenomenon. While a great number of studies have meticulously discerned all the aspects and characteristics of irregular migration within a single national context, the studies that have attempted to systematically compare two or more cases have been practically non-existent. This has had the effect of reducing the possibility to recognize the existing differences.

G. Echeverría, *Towards a Systemic Theory of Irregular Migration*, IMISCOE Research Series, https://doi.org/10.1007/978-3-030-40903-6_5

Following the intuitions suggested by the theoretical framework developed in the first part of the study and the path opened up by a number pioneer works (for instance: Garcés-Mascareñas, 2012; Van Meeteren, 2010; Van Nieuwenhuyze, 2009), the design of the empirical research has been explicitly comparative. As pointed out by Sartori, while every study that uses non-ideographic analytical categories, and therefore refers to some general theory or generalizing framework, is implicitly comparative, "the power of comparison and its usefulness is the highest and the most reliable when it is based on explicit and systematic comparisons" (Sartori, 1991a, p. 27).

Yet, what is the advantage of comparison and why can it be useful to understand more in depth the phenomenon of irregular migration? If there has been a problem of non-differentiation and of "uncontrolled" generalization of the findings gathered in a single context of the whole phenomenon, comparative research can offer an effective remedy. Comparative research allows us to explain because it allows us to control (Sartori, 1991a, 1991b). Only through assessing different cases is it possible to produce law-like statements or, in Sartori's words, "generalizations, with explicative power, that capture regularities" (Sartori, 1991a, p. 27). As put by Garcés-Mascareñas: "Only by comparing and, even more, by comparing what some would call the 'incomparable', is it possible to formulate questions that otherwise would have never been considered and, by so doing, trace relationships and deconstruct categories that are all too often taken for granted in particular historical and national contexts. As Block well puts it: '[C]omparison is a "powerful magic wand" that allowed historians to see beyond local conditions to develop more comprehensive explanations'" (Garcés-Mascareñas, 2012, p. 42).

Then, also for the study of irregular migration, the comparative methodology can be a valuable instrument. The comparative analysis of irregular migration phenomena in different contexts can help to establish differences and similarities, to assess the role of the contextual features in determining specific characteristics, to construct preliminary theoretical frameworks that explain both regularities and peculiarities. This type of approach can help to overcome that lack of theoretical ambition that has been denounced (Bommes, 2012).

Of course, the advantages of comparison are not without a price. "Case studies sacrifice generality to depth and thickness of understanding, indeed to *Verstehen*: one knows more and better about less (less in extensions). Conversely, comparative studies sacrifice understanding-in-context and of context to inclusiveness: one knows less about more" (Sartori, 1991b, p. 253). The important thing, as always occurs when choosing a methodology, is to keep in mind the inevitable, related, trade-offs.

Also concerning the second critique of the current understanding of irregular migration, namely the use of mono-casual explanations, it seems possible to recognize the effects of the mentioned methodological orientation. The vast majority of studies on irregular migration have researched the phenomenon within a single geographical context and using a one-sided theoretical lens, for instance, that of migrant's agency or that of social structures. If this has been the case, it is not surprising that there has been a tendency to produce mono-causal explanations. In particular, as comprehensively discussed in Chap. 3, when the focus has been centred on the role of structures (policies, implementation, the economy, culture, etc.), the role of the migrants' agency and the capacity of migrants to react and adapt to it have been understated. In contrast, when the focus has centred on the migrants' agency (strategies, networks, aspirations, intentions, etc.), the role of structures has been downplayed. Both approaches have missed focusing on "the heart of social life", which is precisely the "interconnections between social agency and systems elements" (Layder, 1998, p. 48).

For this reason, with the awareness of increasing the complexity of the task and the connected risks, it was considered that it was not enough to simply adopt a comparative perspective, but that such comparison needed to include an analysis capable of addressing both social structures and migrants' agency, and their interactions. In this respect, the methodological approach proposed by Derek Layder, which he called "adaptive theory" (Layder, 1998), was of great help and allowed to establish a permanent dialogue between the theoretical and empirical parts of this study. In his perspective:

> "Adaptive theory focuses on the construction of novel theory by utilizing elements of prior theory (general and substantive) in conjunction with theory that emerges from data collection and analysis. It is the interchange and dialogue between prior theory (models, concepts, conceptual clustering) and emergent theory that forms the dynamic of adaptive theory". Moreover, "[M]oving away from empiricism allows the theoretical registering of the systems elements of social life rather than simply those to do with the lifeworld. The empirical focus of the adaptive approach centres on the lifeworld-system linkages that characterize the structure of social reality in general and which are also principal defining features of that area of social life which is currently the research focus".

Following this approach, therefore:

> Both actor's meanings, activities and intentions (lifeworld), and culture, institutions, power, reproduced practices and social relation (system elements) must be taken in account" [...]. The acceptance of both lifeworld and systems features as part of a comprehensive, interconnected and stratified social ontology, also enables a proper treatment of issues of power, control, and domination, and the resources that underpin them (ideologies and cultures). The pervasive influence of power (and control) and the manner in which it manifest itself in different domains of social life cannot be understood properly if its systemic (or structural) aspects are not recognized or registered in the first place (Layder, 1998, p. 48).

With these points in mind, the choice has been to compare the irregular migration phenomenon in two different countries and to use a double research strategy. On the one hand, a context study was developed. This was assembled using secondary literature and the available statistical data and was aimed at assessing the main structural characteristics affecting irregular migration in the two contexts. On the other hand, an original empirical study was developed which aimed at retracing the life experiences of irregular migrants within the selected contexts. The systematic comparison between both the contextual characteristics and concrete experience of irregular migrants in the two countries pointed at establishing similarities and differences and at producing a hypothesis capable of explaining them.

The main research questions that backed the empirical study were: (A) How do the contextual characteristics affecting irregular migration of the two countries differ? In what aspects and to what degree? (B) Are the irregular migration experiences in the two countries different? In what aspects and to what degree? (C) How may the different contextual characteristics affect the irregular migration experiences?

A number of secondary, more concrete, questions guided the research of both contexts and migrants' experiences. Regarding the first aspect, and following the scheme elaborated in the theoretical part of the study, the questions were: (A) What was the migratory history of the country? (B) What have been the main policies affecting irregular migrants? (C) What have been the main characteristics and trends of the economic system and the labour market? (D) What have been the main characteristics of the welfare state? (E) What have been the attitudes of the political and public opinion?

Regarding the second aspect, since the aim was to assess the experience of irregular migrants, four main questions were posed: (A) What has been the legal trajectory (residence and possible regularization) in the host country? What have been the related problems and solutions? (B) What has been the labour trajectory? What problems and what solutions? (C) What was the migrants' experience of internal controls? What problems and solutions? (D) How have other issues, such as healthcare and housing been dealt with?

5.2 Selection of the Cases

As pointed out by Sartori, the choice of the cases to be compared entails a number of problematic issues and possible risks (Sartori, 1991a, 1991b). Obviously, comparison makes sense when there are differences between the selected objects. Yet, similarities are needed too, otherwise the danger is to end up comparing "apples and oranges". When comparing then, the crucial question that needs to be raised is: "comparable with respect to which properties or characteristics and incomparable (i.e. too dissimilar) with respect to which other properties or characteristics?" (Sartori, 1991b, p. 246).

Once the existence of a minimum number of communalities has been established between potential objects of comparison, it is possible to choose between two main strategies (Sartori, 1991b). On the one hand, it is possible to compare "the most likely cases" (Broeders, 2009, p. 20), to use the "most similar system design" (Sartori, 1991b, p. 250). When the objects of comparison are countries, for instance, this means: "choosing for the homogenization of the sample of countries on key aspects considered important" (Broeders, 2009, p. 21). This strategy is especially useful when a particular, very well defined phenomenon that is present in the two contexts needs to be researched. With the contextual similarity, the systemic, contextual features (for these the ceteris paribus criteria is used) can be left in the background and the focus is the analysis of the specificities of the phenomenon. This may be helpful to learn more and to discover further characteristics or to fine-tune an existing theory. The other option is to compare "the most different cases" (Broeders, 2009, p. 20), with the "most different system design" (Sartori, 1991b, p. 250). In this case, the choice of the cases is more disparate. This strategy is particularly useful when the researched phenomenon is more wide-ranging and still not precisely defined. The radical differences between the cases allows us to assess the extension of the researched phenomenon, to control the validity of an early conceptual framework and, most importantly, to observe the effects of the systemic, contextual features on its characteristics (Sartori, 1991b, p. 250).

Given the existing limitations in the theoretical understanding of irregular migration and the interest in assessing the systemic character of the phenomenon, the choice was to adopt the research design of "the most different cases". In particular, what was chosen was to comparatively research the irregular migration phenomenon in the Netherlands and Spain.

The two countries share a number of similarities that validate the possibility of comparison. They are both EU, highly developed, liberal-democratic nation-states. Yet, within this broad group, they also show many important differences, for instance, in their economies, welfare states or social structures. It is especially in relation to the field of our interest, i.e. irregular migration, that the two countries could be considered, somehow, opposite cases.

The Netherlands is an old country of immigration, which has received consistent numbers of migrants since the 1960s. This long experience has translated into a very developed set of policies directed at governing the phenomenon in all its facets. Especially since the 1990s, the efforts by the Dutch government have become increasingly restrictive and today the country has one of the toughest and most efficient policies against irregular migration. As pointed out by Engbersen and Broeders, within "fortress Europe", the Netherlands can be considered as "the heart of the fortress" (Engbersen & Broeders, 2009, p. 870).

Spain, in contrast, is a recent country of emigration, which started receiving consistent numbers of migrants only in the late 1990s. Because of the weak border controls and the recurrent adoption of massive regularization of irregular migrants, like in countries, such as Italy, Greece or Portugal, Spain has been considered as part of the "European soft underbelly" (Pastore, Monzini, & Sciortino, 2006).

As pointed out by Finotelli, the idea of a sharp north/south divide between countries regarding the management of irregular migration should not be uncritically taken, since reality is usually much more complex and nuanced (Finotelli, 2009). Yet, the cases of Spain and the Netherlands can certainly be considered quite different. If the objective was to inquire into the variety and extension of the irregular migration phenomenon, then the comparison of these two cases, precisely because of their differences, appeared particularly stimulating and promising.

If the idea to compare the irregular migration phenomenon within two national contexts seemed promising, the extension and complexity of the task required adopting strategies to reduce the object of analysis. In this respect, there were three main choices.

Firstly, it was decided to focus on one national group of migrants. This allowed to significantly reduce those variables concerning the different origins of the migrants. The selected national group was that of Ecuadorian migrants. There were two main reasons for this choice. The first had to do with the characteristics of the Ecuadorian migration. Though small numbers of migrants had been leaving Ecuador since the 1980s, it was in relation to the deep economic crisis that hit the Andean country at the end of the 1990s, that almost one fifth of the population emigrated in the following decade (Herrera, 2008). At the end of the 2000s, given the economic recovery of the national economy, the migratory trends returned to the pre-crisis standards. With all the necessary caution, then, Ecuadorian emigration can be considered as a relatively time-limited, "one shot" phenomenon, which had basically an economic justification. The second reason had to do with my own Ecuadorian nationality. Besides the obvious personal interest, this fact entailed some potential research advantages. The sharing not only of a common language but also of a number of cultural and communicative codes between the researcher and the people researched, especially in such a sensitive case as that of irregular migrants, may be an element that helps to overcome inevitable barriers and reticence.

Secondly, it was decided to geographically limit the area of consideration to the cities of Amsterdam and Madrid. This not only meant making the fieldwork more feasible in practical terms, but it also allowed for the comparison of two similar settings. Both cities are the biggest urban areas in their countries, they host important immigrant communities, and have developed services and industrial economies.

Thirdly, given the dynamic character of migrants' status and the possibility that both former irregular migrants had regularized or that formerly regular migrants had become irregular, it was decided to interview migrants who had been irregular for at least two or more years during their migratory trajectory.

5.3 Fieldwork Methodology, Strategies and Limitations

The fieldwork research in the cities of Amsterdam and Madrid was mainly developed in 2012/2013/2014.

The fieldwork in Amsterdam was realized between November 2012 and July 2013. In order to develop this part of the research, I moved to Amsterdam for 7 months. I was hosted by an Ecuadorian migrant who had a house in the *Bijlmermeer* neighbourhood in the *Zuidoost* borough of Amsterdam. The fieldwork in Madrid, was realized between August 2013 and February 2014.

The adopted research strategy did not orthodoxly follow any methodological paradigm. On the contrary, it combined a number of strategies and approaches derived from different qualitative methodologies. The main methodological reference, though, was offered by Layder's "adaptive theory" (Layder, 1998). The crucial suggestion of this perspective is to maintain a continuous, bidirectional dialogue between the results of the theoretical reflection and bibliographical analysis, and the results of the empirical research. This flexible strategy allows us to combine both inductive and deductive approaches instead of being limited to just one of them.

Practically, this translates into a process that does not separate the theoretical and empirical phases of the research, or, in metaphorical terms, the library from the street. Instead, the researcher permanently brings the results of his/her readings to the field to test their validity and plausibility and takes the evidence emerged from the field to the library in order to validate, modify or reject the existing theories.

The main research tools used throughout the fieldwork were key informant interviews, participant observation and in-depth interviews with migrants.

5.3.1 Key Informant Interviews

The first step of my fieldwork was the collection of a small number of interviews with key informants. These interviews helped me to establish the general contours of the phenomenon I was going to research. Moreover, they offered a number of indications about possible contacts with migrants and locations where I could encounter them. I collected 6 interviews in the city of Amsterdam (2 with NGO volunteers who help irregular migrants, 2 with spiritual leaders of the Amsterdam catholic church, 1 with the Ecuadorian consul in the Netherlands, 1 with the leader of an Ecuadorian migrants' association) and 5 interviews in the city of Madrid (2 with NGO volunteers, 1 with the Ecuadorian consul in Spain, 2 with leaders of the Ecuadorian migrants' association).

5.3.2 Participant Observation

One of the main objectives of the fieldwork was to directly observe the daily activities of irregular migrants in the two cities. To this end, I adopted two strategies. Firstly, I tried to get invited to and participate in a wide variety of social activities such as: reunions, parties, festivities, religious events, sports gatherings, etc. I have always been clear with the migrants about my aim. I told them that I was a researcher and that I was trying to understand how irregular migrants live, what problems they have and how they solve them. During the events in which I had the chance to participate, the main objective was to observe people's behaviour, to listen to conversations, to collect personal impressions and ideas. These were also opportunities to chat with people, to establish relations and to find potential candidates for the in-depth interviews. At the end of the day, I always elaborated field notes in which I compiled all the collected impressions and information. Then, as I started to develop closer relations with certain migrants, I asked them if I could meet their families, go to their houses or go with them to work. Also in these cases, I had the chance to develop many friendship relations. I always bore in mind my aim and asked for permission if I wanted to use information or to quote a certain conversation in my field notes.

5.3.3 *In Depth-Interviews*

During the first phases of the fieldwork and in particular during the participant observation, I was able to identify possible interesting candidates for in-depth interviews. On those occasions I asked these people if they wanted to take part in my research and I explained my goals and the procedure to them. I generally received enthusiastic replies to my request. The interviews were realized in different places such as the migrants' houses (the majority), bars, parks, public libraries, train stations. The setting and the availability of time on the part of the interviewees had an effect on the length of the interviews. The average length was about 2 h, the shortest was 30 min and the longest 7 h. In Amsterdam I collected 32 interviews, and in Madrid 31.

During the interviews I used a one page general scheme that helped me to keep in mind the main topics I wanted to discuss and some key questions. My aim, however, was to maintain, as much as possible, a conversational, free approach. The idea was to introduce issues and then to allow other topics to emerge and develop along with the flow of the conversation. In many cases, this strategy not only determined a temporal extension of the interviews but also the discussion of topics not necessarily pertaining to the research goals. It is my conviction, nevertheless, that this strategy made it possible to somehow break the rigidity of the interviewer/interviewed roles and therefore to produce a richer exchange and more useful information.

5.3.4 *Study Limitations*

Both the selection of cases and the chosen methodological strategies entail a number of possible problems and limitations that it important to make explicit and to reflect upon.

Regarding the first aspect it seems important to discuss two issues. Firstly, the choice to study the experience of Ecuadorians, and therefore of my co-ethnics, as previously discussed, may certainly offer certain advantages but also some limitations. While the share of a common language and of a common cultural background can be a useful tool for instance to "break the ice" or to better understand certain meanings and expressions during the research, this can also imply many downsides. The presence of culturally structured and codified elements in the relation between the researcher and the researched may importantly influence both sides. For the former, for instance, this may translate into prejudices, preconceptions or "taking for granted" forms of biases; for the latter, into reluctance, hesitancy or the desire to appear in a certain way instead of another. Although, every research relation, i.e. one between co-ethnics or one between people of a different origin, necessarily implies specific problems, what seams important is to keep them in mind, to reflect upon them, and, if possible, to develop strategies to limit their effects. In my case, I tried to combine, on the one hand, mental openness and a questioning attitude towards my subjective impressions and conclusions, on the other hand, discretion and a certain restrain in the expression of my opinions, personal history or social background. The first attitude helped me to limit the role of preconceptions or "giving for granted" assumptions; the second helped to reduce the possible influence of my subjectivity upon the interviewed.

Secondly, when choosing the two countries to study, the option for "the most different cases", as mentioned, has the advantage of offering a great variance, which can be useful to search for the extension of a phenomenon. Yet, a connected risk to this strategy is that the enormous difference between the cases may end up resulting in an unproductive comparison. What it was attempted, to avoid such risks is to select two cases that, being very different, present yet sufficient commonalities in order to allow a fruitful contrast.

Regarding the chosen methodologies and in particular the option for in-depth interviews and ethnography, as more in general occurs with qualitative approaches, a number of issues may be raised concerning the validity of the collected data. Firstly, it is important to remember the lack of statistical validity given the limitation of the sample and the research techniques. Secondly, the relay on personal assessments, memories and anecdotes, the bulk of the interviews material, implies the potential interference subjective not only psychological but also "environmental" distortive elements. These two important limitations require a pondered use of the possible findings. In particular, on the one hand, given the lack of statistical validity, the elements that emerge from the fieldwork must be used not as definitive indicators or as conclusive demonstrations but rather as elements able to suggests hypothesis, contest convictions, open interpretative possibilities. On the other hand, the use of "personal material", requires an extra effort of material analysis, comparison and cross-check both between the different interviews but also with other research material such as statistics, previous studies, etc.

Bibliography

Bommes, M. (2012). Illegal migration in modern society: Consequences and problems of national European migration policies. In C. Boswell & G. D'Amato (Eds.), *Immigration and social systems. Collected essays of Michael Bommes* (pp. 157–176). Amsterdam: Amsterdam University Press.

Broeders, D. (2009). *Breaking down anonymity digital surveillance on irregular migrants in Germany and the Netherlands*. Rotterdam, The Netherlands: Erasmus Universiteit.

Engbersen, G., & Broeders, D. (2009). The state versus the Alien: Immigration control and strategies of irregular immigrants. *West European Politics, 32*(5), 867–885. https://doi.org/10.1080/01402380903064713

Finotelli, C. (2009). The north–south myth revised: A comparison of the Italian and German migration regimes. *West European Politics, 32*(5), 886–903. https://doi.org/10.1080/01402380903064747

Garcés-Mascareñas, B. (2012). *Labour migration in Malaysia and Spain: Markets, citizenship and rights*. Amsterdam: Amsterdam University Press.

Herrera, G. (Ed.). (2008). *Ecuador: la migración en cifras*. Quito, Ecuador: FLACSO-UNFPA.

Layder, D. (1998). *Sociological practice: Linking theory and social research*. London: Sage.

Pastore, F., Monzini, P., & Sciortino, G. (2006). Schengen's soft underbelly? Irregular migration and human smuggling across land and sea borders to Italy. *International Migration, 44*(4), 95–119.

Sartori, G. (1991a). Comparazione e metodo comparato. In G. Sartori & L. Morlino (Eds.), *La comparazione nelle scienze sociali* (pp. 25–45). Bologna, Italy: Il Mulino.

Sartori, G. (1991b). Comparing and miscomparing. *Journal of Theoretical Politics, 3*(3), 243–257.

Van Meeteren, M. (2010). *Life without papers: Aspirations, incorporation and transnational activities of irregular migrants in the Low Countries*. Rotterdam, The Netherlands: Erasmus Universiteit.

Van Nieuwenhuyze, I. (2009). *Getting by in Europe's urban labour markets: Senegambian migrants' strategies for survival, documentation and mobility*. Amsterdam: Amsterdam University Press.

Chapter 6
Ecuadorian Migration in Amsterdam and Madrid: The Structural Contexts

The scope of this chapter is to outline the main characteristics of the two contexts that were the scenario of the phenomenon that is the object of this book. The analysis will centre on those structural features of the cities of Amsterdam and Madrid and, more in general, of the Netherlands and Spain, that may have had an influence on the experience of Ecuadorian irregular migrants.

The chapter will be divided into two parts. In the first, the main features of the Ecuadorian emigration phenomenon will be described. In the second part, a comparative analysis of the structural characteristics of the two receiving contexts will be presented. Following the systemic approach introduced in the first part of this study, the attempt will be to sketch, although in very general terms, some of the features exhibited by the different social systems in Amsterdam and Madrid. To accomplish this goal, a more general discussion about the Netherlands and Spain will be presented and this will focus on five areas that are especially significant to the irregular migration phenomenon: A. migration history and contemporary trends; B. the migration regime[1]; C. the economy and labour market; D. the welfare state; E. the political and public opinion in relation to migration.

Since the Ecuadorian emigration phenomenon has followed a characteristic temporal pattern, with massive outflows condensing between 1999 and 2006, our analysis of the two destination contexts will focus on the period 1998–2013. The migrants interviewed during the fieldwork, realized in 2013, were all part of the mentioned flux, with few limited exceptions.

Although for both Amsterdam and Madrid, and more in general for the Netherlands and Spain, a vast and extremely valuable literature is available on migration and specifically on irregular migration, in this chapter there will be not a systematic discussion of it. This choice does not mean underestimating the importance of the previous works and their results but, rather, it is intended as part of a strategy aimed at limiting, as much as possible, the introduction of "external", pre-constructed inter-

[1] For the concept of migration regime see (Cvajner, Echeverría, & Sciortino, 2018).

G. Echeverría, *Towards a Systemic Theory of Irregular Migration*, IMISCOE Research Series, https://doi.org/10.1007/978-3-030-40903-6_6

pretative frameworks at this stage. The aim is to allow a more spontaneous and "in-mediate" analysis of the relation between the structural contexts and the results of the empirical research presented in Chap. 7. For this reason, while references to the existing literature will be offered, the discussion will focus mainly on data and figures offered by datasets, official reports and empirical research. Regarding the use of statistical data, given the ample variety of sources available, what was chosen was to privilege the international sources (OECD, Eurostat, World Bank) over the national ones (*Centraal Bureau voor de Statistiek* and *Instituto Nacional de Estadística*), when not possible otherwise. Although this option has some disadvantages, related to the fact that national statistics are usually more precise and disaggregated, the advantages lie in the easier and more direct comparability of the international data, a crucial aspect for a comparative research endeavour. This notwithstanding, it is important to be keep in mind the mentioned weaknesses and assume a dose of caution when proposing conclusions built upon this type of data.

6.1 Ecuadorian Emigration

Although Ecuadorian emigration has been going on in small numbers since the 1970s, the phenomenon reached massive proportions at the end of the past-century. After 1999, and within a matter of a few years, almost an eighth of the entire population (Herrera, 2008; Herrera, Moncayo, & Escobar, 2012; INEC, 2013) left the country in search of a better future abroad. This dramatic change in the migratory pattern of the country was mostly determined by the serious economic and financial crisis that hit the country and culminated with the freeze of private bank accounts in 1999 and the dollarization of the economy in 2000. These outcomes were the result of a long-term process of social and political conflict characterized by corruption, economic inefficiency and the slow but continuous erosion of the political system (Acosta, 1998; Echeverría, 1997; Ramírez & Ramírez, 2005).

Probably the most noticeable effect of the systemic crisis was precisely the sudden and massive migratory outflow. Until 1998, emigration had been relatively limited and registered numbers that were inferior to a thousand per year. Things changed in 1999 when the flux reached hundreds of thousands (Boccagni, 2007). From this moment on, and for the next decade, the outflows presented unprecedented figures (see Fig. 6.1). In the years 2000 and 2002 fluxes peaked above 150,000 people per year. The remittances sent by migrants soon became the second source of national income, passing from 794 million USD in 1998 to 2318 in 2005 (Herrera, 2007; Herrera et al., 2012). The magnitude of the phenomenon changed the social and political understanding of migration; those that once had been considered betrayers started to be considered heroes. The expatriate community was designated officially as the Fifth Region of the country (in addition to the traditional four) and its participation in domestic political life was strongly encouraged (G. Echeverría, 2014a).

The three most important destinations of Ecuadorian migration were Spain, the United States and Italy (Herrera, 2008). However, Spain received by far the largest

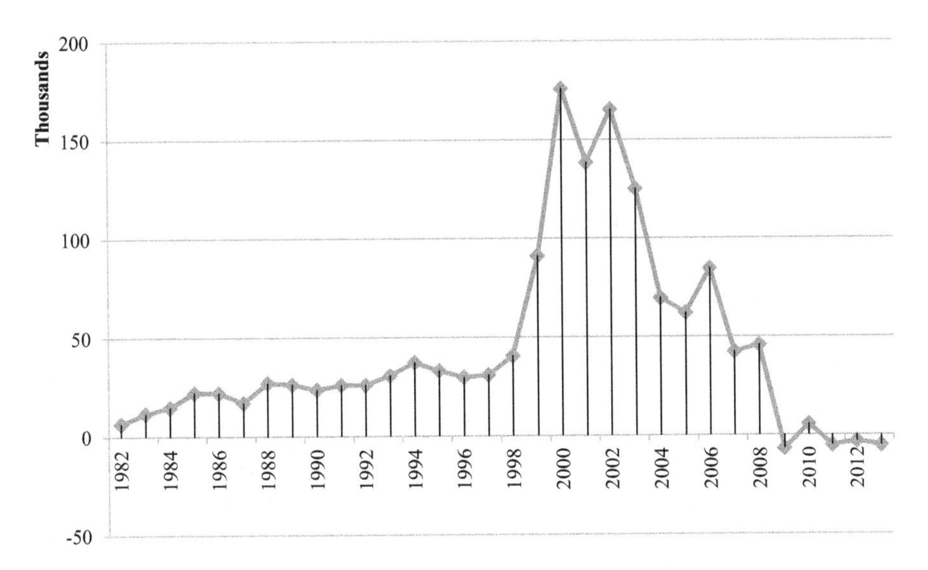

Fig. 6.1 Ecuadorian emigration (Data from: FLACSO-UNFPA, 2008 and INEC, 2013)

part of the flux. This has been related to a number of factors, such as: the common language, the cultural affinity, the visa-free entry and the booming economy at destination (Gómez Ciriano, Tornos Cubillo, & Colectivo IOE, 2007; Herrera et al., 2012). In 2005, Ecuadorian-born immigrants living in Spain reached a peak of 487,239 but slightly decreased in the following years (Eurostat). The other European countries received smaller numbers of Ecuadorian migrants. The Netherlands reached a peak of 3028 Ecuadorian-born people in 2014 (Eurostat).

It was only in 2007 and 2008, as a result of both the improved economic conditions in Ecuador and the beginning of the economic crisis in the US and Europe, that the fluxes went back to the pre-crisis standards. In 2009, and in the following years, since a return-migration phenomenon emerged, net migration registered negative values for the first time in recent Ecuadorian history. The magnitude of these flows, however, never reached the level of those of the previous phase. Although a growing number of those who had left considered the option to return, a large majority decided to remain abroad (Herrera et al., 2012).

6.2 The Netherlands as Irregular Migration Context

6.2.1 Migration History and Contemporary Trends

After the end of WWII and for the next decade, the Netherlands was a country of emigration. This pattern radically changed in the early 1960s. From that moment on, and regardless of the self-perception of its political leaders, which continued to officially refuse that reality until the 1990s, the country has constantly been an important migration destiny.

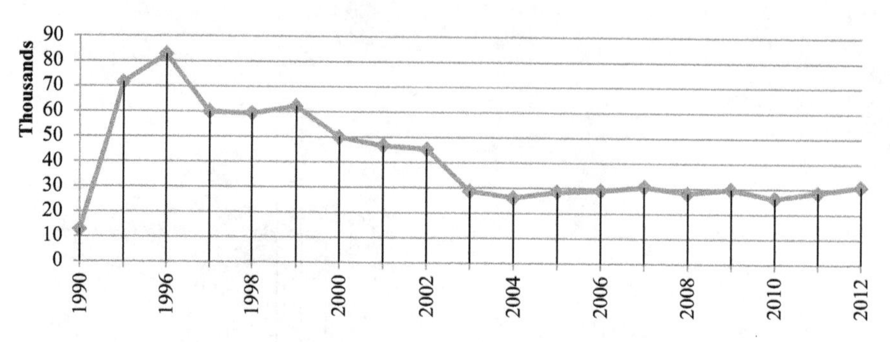

Fig. 6.2 Asylum seeker requests (Eurostat)

Migration researchers have identified a number of important, successive migratory waves in the recent history of the Netherlands (Broeders, 2009; Leerkes, 2009; Lucassen, 2001; Van Meeteren, Van de Pol, Dekker, Engbersen, & Snel, 2013). The first one took place between the early 1960s and the oil crisis of 1973. This wave involved labour migrants arriving in the Netherlands as guest workers from Mediterranean countries, such as Spain, Italy, Portugal, Turkey and Morocco. In the intentions of the Dutch government, these migrants were expected to stay only temporarily and leave the country once their labour contracts had ended. Yet, this plan turned out to be wrong; the vast majority of migrants, and especially those from Turkey and Morocco, decided to stay, establishing significant communities in the main cities.

The second migratory wave was very much related to the first one. Contrary to political prediction, not only did former guest workers not leave after the recruitment ban but, thanks to the existing legal framework, they were able to bring their families to the Netherlands. Moreover, as second generations started to develop, many young males brought spouses from their origin countries. Since the mid-1970s and until our days, these channels have allowed a continuous flux of new emigrants, especially from Turkey and Morocco.

A third important migratory wave involved migrants arriving from former Dutch colonies. These fluxes started in 1975 after the independence of Suriname. In the following years, almost one third of the entire population left the South American country. Furthermore, in the late 1980s, a new stream of immigrants started to arrive in the Netherlands from the Dutch Antilles.

A fourth wave of immigration emerged in the late 1980s and involved asylum seekers (see Fig. 6.2). This flux became particularly relevant in the 1990s when a number of wars and humanitarian crises in Europe and in neighbouring areas determined a sharp increase in asylum requests. Given the generosity of the existing legal framework, the Dutch government had its hands tied once a request was issued and the rate of acceptance was very high. This phenomenon contributed to the so-called "migration crisis" of the 1990s and a political backlash that fostered a serious revision of the migratory and asylum regime in the years to follow.

Finally, a fifth wave of immigration emerged in the 2000s. This flux involved mainly labour migrants from Western countries and in particular from Eastern European countries such as Poland, Bulgaria and Romania.

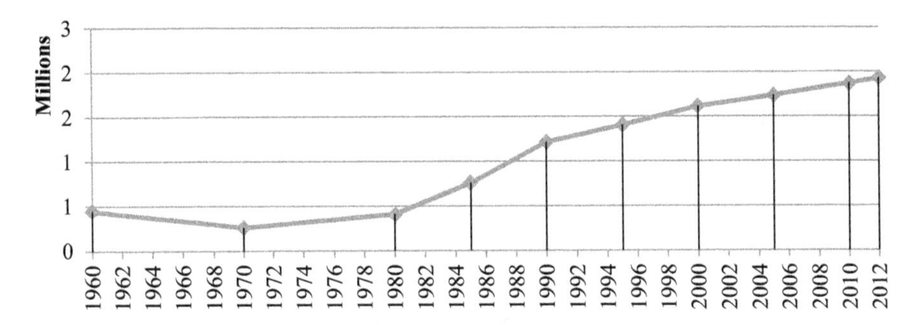

Fig. 6.3 Foreign-born population. (Data from: OECD)

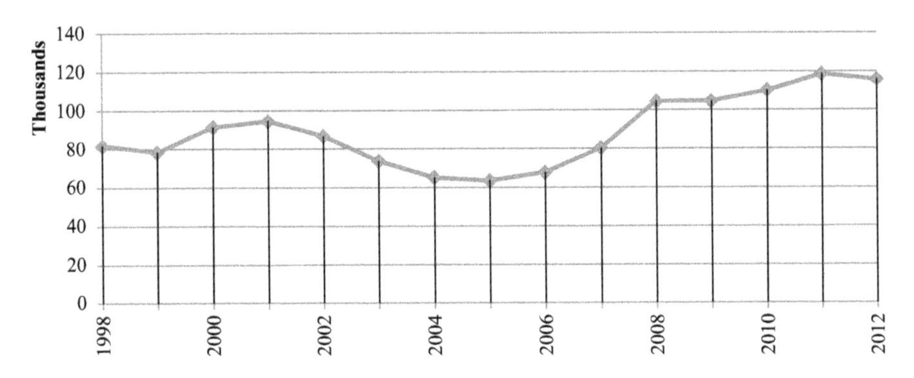

Fig. 6.4 Annual migration inflow. (Data from: OECD)

In 2012, of a total population of 16,730,348 individuals, the foreign-born population in the Netherlands counted 1,927,700 individuals, which represented 11.5% (OECD). The population with a foreign background counted 3,494,193 individuals and represented 20.9% (CBS). This outcome was the result of more than 50 years of continued migration inflows (see Fig. 6.3).

As regards the yearly inflow of new migrants during the 1999–2012 lapse (see Fig. 6.4), a mixed picture emerges. Between 1998 and 2001, the fluxes slightly grew to reach almost 100,000 new arrivals in 2001; from 2001 to 2005, a significant decrease was observable, with a minimum of 60,000 new entries in 2005; from 2005 on, fluxes started to grow again and reached a maximum of almost 120,000 in 2011.

6.2.2 *Irregular Migration Estimations*

A number of estimations of irregular migrants residing in the Netherlands have been produced in the last decade (Engbersen et al., 2002; Hoogteijling, 2002; Leerkes, van San, Engbersen, Cruijff, & van der Heijden, 2004; van der Heijden, Cruyff, &

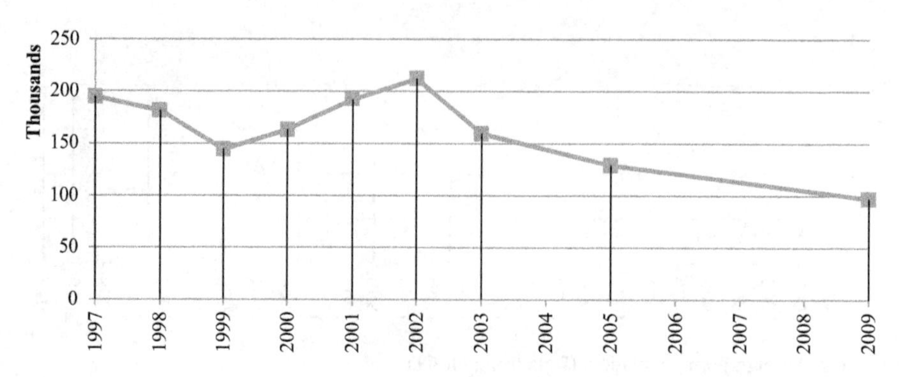

Fig. 6.5 Irregular migration estimation. (Data from: INDIAC – NL EMN NCP (2012)

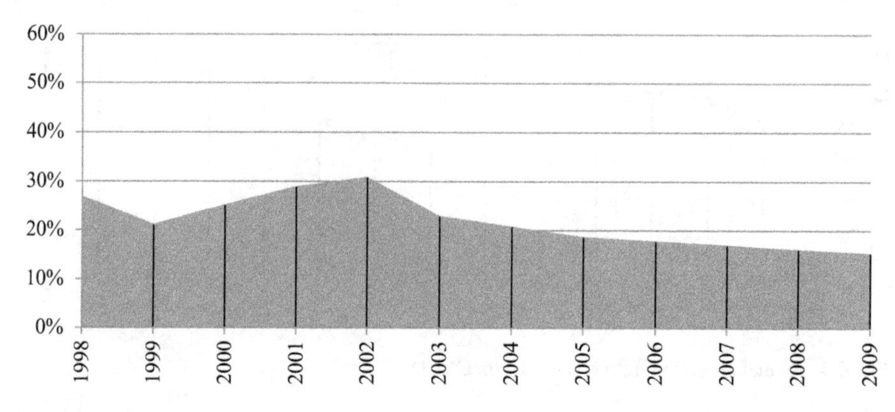

Fig. 6.6 Irregular-migrant percentage of total foreign population (Own elaboration, data from: INDIAC – NL EMN NCP (2012) and Eurostat

van Gils, 2011; van der Heijden, Gils, Cruijff, & Hessen, 2006). Van der Leun and Illes have comprehensively discussed the different methodologies and approximations used by researchers, as well as the main pros and cons of their works. As they pointed out: "methods have been fine-tuned and the quality of available data has gradually improved as a result of increased co-ordination between different government branches and on-going computerization" (Van Der Leun & Ilies, 2008, p. 13).

As can be observed (see Figs. 6.5 and 6.6), the first data available estimated a population of approximately 200,000 irregular migrants in 1997. This figure represented more than 25% of the total foreign population. In the next 2 years, numbers slightly fell to start growing again in the year 2000. The rising trend lasted for the next 2 years. In 2002, the irregular-migrant population in the Netherlands reached a historic maximum of 211,990 individuals (INDIAC – NL EMN NCP & Diepenhorst, 2012, p. 83), which represented more that 30% of the foreign population. Since that year, an opposite and prolonged decreasing trend has been registered.

As underlined by the 2012 Report, *Practical measures for reducing irregular migration in the Netherlands*, an important part of the explanation for this reduction is related to the European Union's enlargements in 2004 and 2007, which determined the automatic regularization of Bulgarian and Rumanian citizens (INDIAC – NL EMN NCP & Diepenhorst, 2012, p. 84). The last available data show for the year 2009 an irregular-migrant population of nearly 100,000 individuals, which represented 15% of the total foreign population.

As summarized by Leerkes, a number of general features characterize the irregular population in the Netherlands. The phenomenon is concentrated in certain agricultural areas and in deprived urban neighbourhoods where irregularity can reach 6% or 8%; irregular migrants originate from more than 200 countries and the largest groups are Turks, Moroccans, Algerians and Surinamese; refused asylum seekers are estimated to constitute 15% of the irregular population (Leerkes, 2009, p. 16).

6.2.3 Migration Regime

Migration scholars have distinguished three phases in the ways in which the Dutch society has dealt with the arrival and residence of irregular migrants (Broeders, 2009; Engbersen, 2001).

The first, corresponding to the decade of the 1960s, was characterized by the "welcoming of 'spontaneous migrants' who could easily be legalized and employed in factory work and agriculture" (Engbersen, 2001, p. 241). The second phase, from 1970 to 1991, was that of "the silent toleration of 'illegal workers', which enabled them to gain access to the formal labour market and take care of themselves" (Engbersen, 2001, p. 241). Irregular migrants during those years "were duly regularized as they found a job" (Kloosterman, Van Der Leun, & Rath, 1999). Broeders has described this phase as characterized by the application of the traditional Dutch principle of *gedogen* (Broeders, 2009, p. 63). This principle, of which he proposes a translation into English using the term toleration, implied an intentionally weak application of the formal legal framework.

> Irregular migrants, once established, are able to find work even in the formal labour market. They can still obtain Social-Fiscal numbers (so-called SoFi numbers), which allow them to hold tax-paying jobs. The enforcement regime on irregular labour is lax and in a number of sectors such as agriculture and horticulture, where despite the high unemployment figures employers find it difficult to fill the vacancies, the authorities often turned a blind eye (Broeders, 2009, p. 63)

The third phase, which started in 1991 and is currently on-going, has been characterized by a radical change in the political and legal approach towards irregular migration. Engbersen has summarized the new paradigm as directed at "excluding and deporting 'illegal aliens'" (Engbersen, 2001, p. 241). A number of consecutive legal reforms and new administrative regulations have been approved with the objective of reducing the irregular migration population. Several research works have analysed the scope, evolution and consequences of these interventions

(Broeders, 2009; Broeders & Engbersen, 2007; Engbersen & Broeders, 2009; Engbersen & Van Der Leun, 2001; Engbersen et al., 2004; Kloosterman et al., 1999; Leerkes, 2009, 2016; Van Der Leun, 2003, 2006; Van der Leun & Bouter, 2015; Van Der Leun & Ilies, 2008; Van Meeteren, 2010; Van Meeteren et al., 2013).

The strategy adopted by the Dutch government has been threefold (see for instance: Broeders, 2009; Leerkes, 2009). A first group of measures had the objective of limiting the entry of new irregular migrants. Crucial actions in this area, often adopted in coordination with the European Union partners, have been: A. the enforcement of stronger and more sophisticated border control systems; B. the tightening of visa policy both for tourist and workers (tougher conditions, extension of the list of countries with visa obligation); C. the limitation of family reunification and stricter marriage policies; D. the fight against human trafficking; E. the sharpening of asylum policy; F. the adoption of limited regularization processes directed towards long-term asylum seekers.

A second group of interventions has focused on making residence for irregular migrants more difficult and costly. The two pillars of this policy were: the exclusion of irregular migrants from important institutions of the welfare state and the fight against irregular employment. As regards the first objective, the most important step was taken with the adoption of the Linking Act (*Koppelingswet*) in 1998. This provision established a link between the possibility to access public services, such as, social security, health care, education or public housing, and the holding of a valid residence permit. Concerning the second objective, numerous actions have been adopted since the early 1990s, for instance: A. the denial of social security and tax numbers to irregular migrants; B. the obligation for employers to check employees' documentation; C. the increase of fines for dishonest employers; D. the allocation of more resources and personnel to the labour inspection service. The implementation of all these policies required fundamental and recurrent improvements to the database and information exchange systems at all the administration levels.

A third group of policies was aimed at making the apprehension, identification and expulsion of irregular migrants more efficient. The actions taken to achieve this goal included: A. tighter policy on individuals' identification obligation; B. stricter controls of employment places; C. the implementation of sophisticated identification technologies; D. improvement of the detention policy (new facilities and longer detention times); E. readmission agreements with third countries; F. improvement of database and information exchange systems at a European level.

6.2.4 Economics, Labour Market and Underground Economy

As can be observed in Fig. 6.7, the Dutch Gross Domestic Product (GDP), used here as a general indicator of the economic trends, shows a fluctuating picture within the considered lapse of years. Between 1998 and 2001, the economy markedly grew, with a peak in the year 1999 with an over 4% variation. Years 2002 and 2003 were characterized by stagnation. In the next 5 years, until 2008, the economy grew

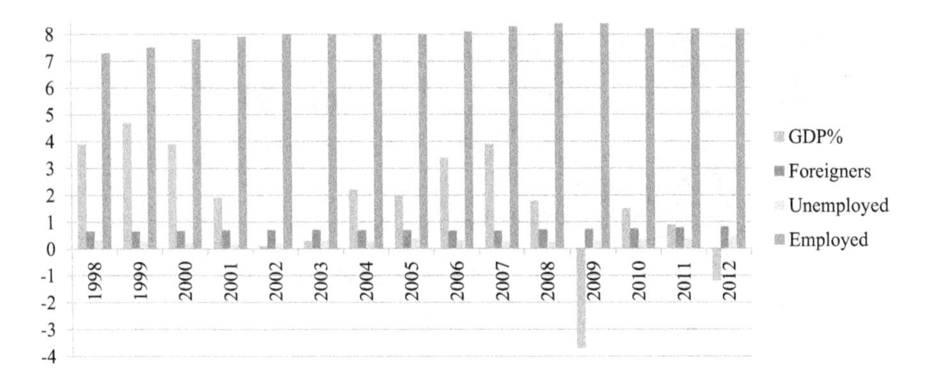

Fig. 6.7 GDP variation, employment, unemployment, migration. (Data from: Eurostat. Foreigners, Employed and Unemployed are in millions per year)

again, especially in 2006 and 2007 when the variation was over 3%. The effects of the economic crisis hit the Dutch economy severely in the year 2009, when the GDP registered a − 3.7% fall. A slight recovery was observable in 2010 and 2011, when the GDP averaged a 1% annual growth. Nevertheless, the economy contracted again in 2012, registering a − 1.2% variation.

Concerning the labour market and in particular total employment, a growing trend has been observable. In the year 1998, the number of employed people was 7,347,100. After 14 years, in 2012, the number rose to 8,254,100. The number of jobs created in this lapse of time was 907,000. It is possible to witness a direct, however slightly delayed, correlation between the GDP and the jobs created. The years when the economy grew were those when also the labour market expanded. On the contrary, a contraction of the GDP, like the one that occurred in 2009, determined a significant destruction of jobs. A year later, in 2010, the labour market had lost 216,500 jobs.

As regards the unemployment, an inverse, slightly delayed, correlation with the GDP has been observable. In general (see Fig. 6.8), very low numbers have been registered. The peak was reached in 2012, when 460,000 people were unemployed; they represented 5.57% of the active population.

With regard to the occupation structure (see Fig. 6.9), the Dutch labour market did not undergo noteworthy changes in the considered years. In 1998, highly skilled occupations (International Standard Classification of Occupations – ISCO are used), such as Managers, Professionals, and Technicians and Associate Professionals, represented 46%; 14 years later, the same group had fallen one percentage point to 45%. In the same years, Elementary Occupations, passed from 7% to 8.3% in 2012.

Finally, as regards the underground economy (see Fig. 6.10), the estimations produced by Schneider and his colleagues for the Netherlands, evidence a decreasing trend in the considered years (Schneider et al., 2010; Schneider, Raczkowski, Mróz, & Futter, 2015). In 1999, the underground economy represented 13.3%; in 2014 it had fallen to 9.2%. Both percentages are way below the European Union (28 countries) average, which scored a 20.3% in 1999 and 18.6% in 2014.

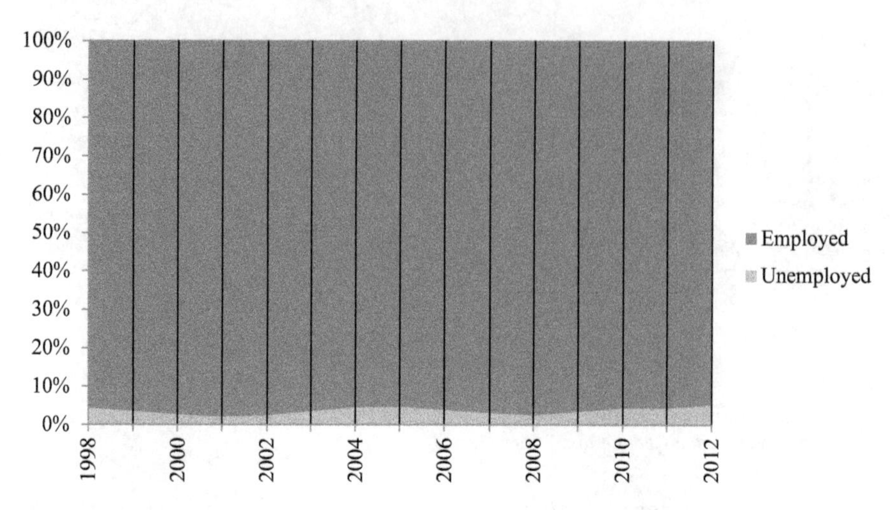

Fig. 6.8 Labour market structure. (Data from: Eurostat. Foreigners, Employed and Unemployed are in millions per year)

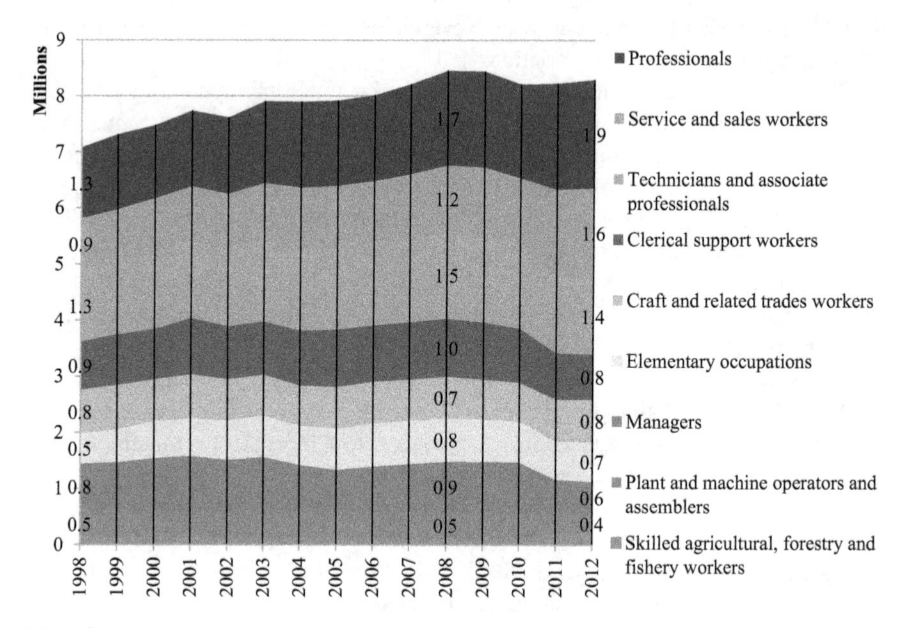

Fig. 6.9 Main occupations. (Data from: Eurostat)

6.2.5 *The Welfare Regime in the Netherlands*

Within the different welfare state clusters in Europe, the Dutch welfare state is usually placed under the heading of the so-called Continental Welfare Regimes (Esping-Andersen, 1990, 1996b; Ferrera et al., 2000; Hemerijck, 2012; Hemerijck, Keune, & Rhodes, 2006; Hemerijck, Palm, Entenmann, & Van Hooren, 2013). Hence, its

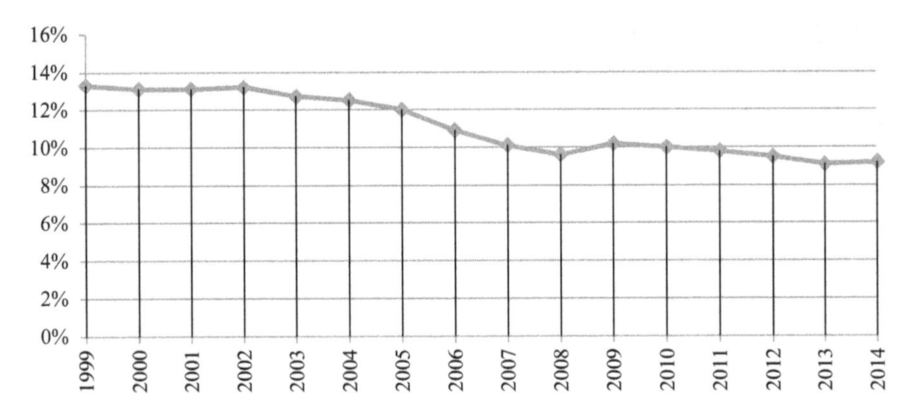

Fig. 6.10 Underground Economy. (Data from: Schneider et al., 2010, 2015)

original conception was based on the Bismarckian tradition. Following the social insurance model, "a tight link between work position and/or family status and social entitlements" (Hemerijck et al., 2013, p. 21) was established. As pointed out by Hemerijck and his colleagues, the influence of the Christian tradition and, in particular, of Calvinism was strong behind this conception. "The Calvinist emphasis on individual responsibility makes Calvinism rather suspicious of establishing poor relief programs without enforcing work discipline, next to only meagre relief" (Hemerijck et al., 2013, p. 21).

The insurance model was functional to "the status maintenance and the support of traditional male breadwinner nuclear family structures" (Hemerijck et al., 2013, p. 21). Accordingly, the labour market was strongly regulated and focused on enabling the possibility of long, stable, remunerative careers. Women were discouraged from participating in the labour market, they received "indirect social protection though derived male breadwinner stable employment, social insurance and passive familiar benefits" (Hemerijck et al., 2013, p. 21). Within this model, those who were unable to follow the job-insurance path had to rely on a network of local social assistance organizations.

While the Bismarckian tradition was at the base of the Dutch welfare state, a number of features indicated a certain distance from orthodoxy. In particular, the provision of basic public pension, the tax-financed minimum social assistance and the public financing of elderly care services clearly signalled a departure from a strict insurance model (Hemerijck et al., 2013, pp. 21–22).

The Dutch welfare state has undergone a process of radical reforms since the 1980s and increasingly in the 1990s and 2000s. These reforms implied "an explicit U-turn away from the Continental pathology of "welfare without work" towards embracing a more inclusive and activating welfare state" (Hemerijck et al., 2013, p. 27). Similarly to what occurred in the rest of Europe, welfare state recalibration was largely motivated by the deep and complex structural changes affecting societies and states (Esping-Andersen, 1996a; Ferrera, 2008; Ferrera et al., 2000;

Hemerijck, 2012). Within the new demographic, productive and competitive conditions, the very sustainability of the welfare state was at stake.

The changes introduced by the successive reforms in the Netherlands implied the gradual move away from Bismarckian employment-related social insurance towards a basic universal income support based on general taxation. "Fighting poverty has become a new distributive priority, that implied a shift in attention from insiders (male breadwinners, their dependents and societal representatives) to outsiders (women, low-skill groups and others)" (Hemerijck et al., 2013, p. 28). This shift was complemented by a comprehensive reform of the labour market policy. The emphasis in this area was now placed on the activation and increasing insertion of previously- excluded sectors of the population (women, the elderly the unemployed, low-skilled workers and migrants) in the labour market. The new paradigm was captured by the concept of "flexicurity". The agreement between the government, the trade unions and the employers, in 1995, allowed a flexibilization and diversification of the types of contracts in exchange for a universal protection system. Successive reforms (2000, 2002) further extended the labour rights and protections connected to flexible contracts, leading to a *de facto* equalization with those granted by permanent contracts. In the field of activation, a number of measures were taken through the late 1990s and 2000s, and the objective was to incentivize work at all levels. The measures included: A. the implementation of counselling and permanent training systems for the unemployed; B. the discouragement of early pensions and the reduction of disability benefits; C. the implementation of policies to reconcile work and family life through parental leave incentives, subsidies, tax deductions.

These important transformations of the Dutch welfare state required "strengthening the role of the central government and local authorities, at the expense of the social partners" (Hemerijck et al., 2013, p. 28). Moreover, both the promotion of active labour market policies and the development of more sophisticated systems to provide social services required a continuous modernization of the administrative apparatus. All in all, as pointed out by Broeders, the Netherlands has an "elaborate welfare state with a high level of social protection, which requires a keen eye for matters of eligibility. Most sectors of public and semi-public life are highly regulated and subject to registration and documentary requirements by a professional and well-staffed bureaucracy" (Broeders, 2009, p. 40).

6.2.6 *Politics, Public Opinion, Migration*

After three decades of sustained migrations and the development of important communities of emigrants in the main cities, in the early 1990s the Dutch government still refused to officially recognize the Netherlands as a country of immigration (Van Meeteren et al., 2013). A historic step was taken in 1998, when the role of migration was officially acknowledged as central to the Dutch society. Yet, this step, which caused heated debates in the Parliament, was nothing more than an act of self-conscience or self-recognition.

While the reality of immigration had been officially understated, the Dutch government had been actively dealing with it since the 1970s. In those years migration had been generally welcomed. As pointed out by Kloosterman and his colleagues: "only three decades ago, the Dutch government welcomed undocumented immigrants who were represented as 'spontaneous guestworker'. They were duly regularized as soon as they found a job" (Kloosterman et al., 1999, p. 252). As regards the integration of the newcomers, a multiculturalist approach was adopted (Entzinger, 2006; Van Meeteren et al., 2013). Migrants should integrate while preserving their ethnic identity: "the emphasis was on self-organization and arrangements for education in minorities' own languages. […] The immigrant integration policy aimed at mutual adaptation and equal opportunities for Dutch people and ethnic minorities" (Van Meeteren et al., 2013, p. 118).

The multiculturalist perspective became criticized in the 1990s. Migrant communities showed significantly higher levels of unemployment, welfare dependency and marginalization. The new approach, then, focused on the socio-economic integration of migrants. "Integration was interpreted as equal participation in the major social institutions of society" (Van Meeteren et al., 2013, p. 119).

In the early 2000s, the government's approach towards migrations underwent another transformation. While the political and social attitude towards migration had been deteriorating since the 1990s, in connection to the increasingly conflictual relations with the immigrant communities and the sustained arrival of new flows, the first years of the new millennium meant a turning point. On the one hand, a number of dramatic episodes at a national and international level, for instance, the assassination of Pim Fortuyn (2002) and Theo Van Gogh (2004) or the 9/11 terrorist attacks, raised the alarm about the effective integration and possible "integrability" of the immigrant communities and especially of those of Muslim religion. On the other, populist Dutch Politicians, such as Pim Fortuyn, Ayaan Hirsi Ali or Geert Wilders, cleverly exploited these events to support their claims. Slogans like: "the Netherlands is full" and "multiculturalism has failed" became part of a heated public debate (Garcés-Mascareñas, 2011; Penninx, 2006; Van Meeteren et al., 2013).

The changed climate transformed into political action. The new emphasis of integration policies was centred "on the individual responsibility […]. Integration policies became not only strongly related to issues such as shared norms about the rule of law and the obligation to know the Dutch language and culture, but also to social problems of public order and crime. Integration policies became more assimilistic and immigration policies more selective" (Van Meeteren et al., 2013, p. 119). The main policy tool within the "new course" has been the civic integration tests. Although these tests had already begun in 1998, a number of successive modifications (2006, 2007, 2008) extended their scope and considerably increased their difficulty. Migrants willing to travel to the Netherlands for family reunification, family formation (marriage), labour or other reasons, were obliged to pass a paid test in the Dutch embassy of their countries; a minimum knowledge of the Dutch language and Dutch society were necessary. With the 2007 modification, the same requirements were extended to migrants already in the national territory. They had to pay for their own integration courses and were given a certain time to pass the tests. In case of failure, administrative fines were applicable. As has been pointed out, these tests have become powerful tools to restrict migration (Garcés-Mascareñas, 2011).

6.3 Spain as an Irregular Migration Context

6.3.1 Migration History and Contemporary Trends

The transformation of Spain into an immigration country took place in the mid-1980s, after centuries of emigration history. This important event passed somewhat unnoticed by the public opinion and the government in those years (Izquierdo, 1996). When the Spanish government had to negotiate the conditions to join the European Union, a major concern at the bargaining table was the risk of a heavy outflow of workers towards the richer partners of the North. For this reason, the final agreement included a transitory norm that limited the circulation of Spaniards for some years. Contrary to all expectation, the entry of Spain into the European Union, on the first of January, 1986, did not mean an increase in emigration. Ironically enough, it was that year that the net flows changed sign and the inflows surpassed the outflows.

From that moment on, and for the next decade, Spain would experience a slow but continuous increase in migration numbers. These, nevertheless, would be far lower than those experienced by traditional European migration countries (Arango, 2010). It was in the last years of the past century, and especially in the first of the new one, that migration to Spain reached truly spectacular volumes, determining a radical and far-reaching change to the demographic structure of the receiving society.

The arrival of migrants was mainly sustained by a powerful demand for foreign workers which was itself determined by the booming economy (Aja & Arango, 2006; Arango, 2005, 2010; Cachón, 2009; Cebolla & González Ferrer, 2008). Although unemployment among nationals was not marginal, especially among young people, the segmented character of the labour market permitted a complementary integration of the newcomers. These were especially required in a number of specific sectors, in particular: construction, services, agriculture and personal services.

As regards the origin of migrants, the main fluxes arrived from East-Europe (Romania and Bulgaria), Latin America (Ecuador, Bolivia) and North and West Africa (Morocco). The main entry channels were visa overstaying and irregular border crossing. Asylum-seeker requests played a secondary role in comparison to other European countries (see Fig. 6.11) (González-Enríquez, 2009). Within the considered time lapse, the peak was reached in 2001, with 9489 requests. The years to follow, with the partial exception of 2007, saw permanent decrease.

In 2012, of a total population of 46,818,219 individuals, the foreign-born population in Spain counted 6,618,200 individuals, which represented 14.3% (OECD). In

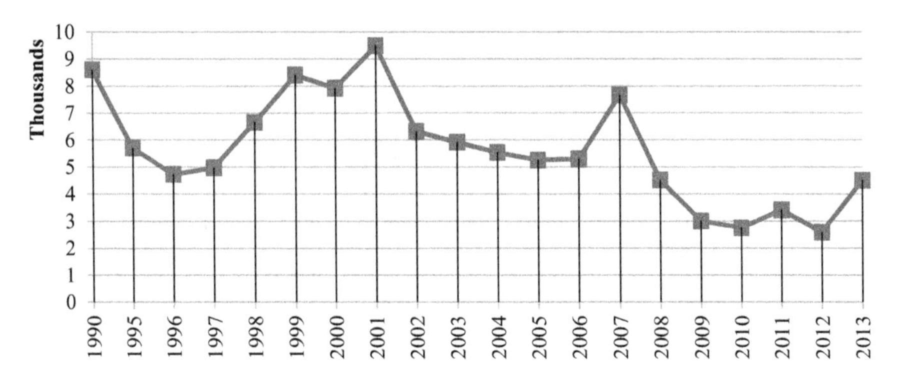

Fig. 6.11 Asylum-seeker requests. (Data from: OECD)

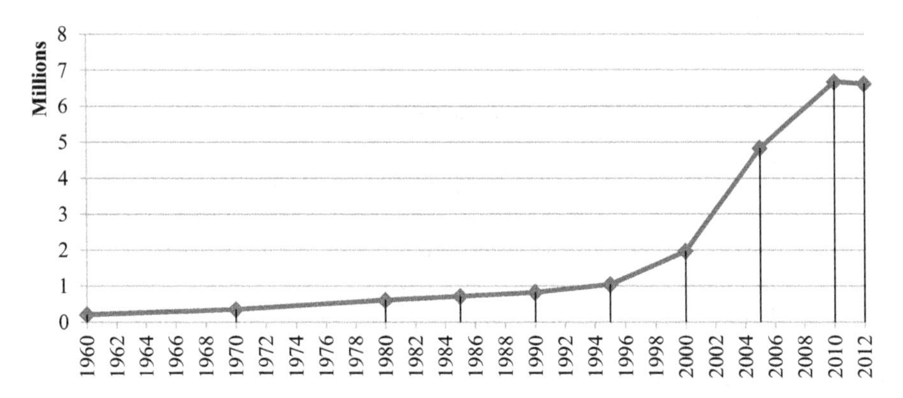

Fig. 6.12 Foreign-born population. (Data from: OECD)

order to have an idea of how fast and sharp the demographic change was, all one has to do is to recall that in 1995, less than 20 years before, the foreign-born population counted 1,401,200 individuals which represented 2.6% (see Fig. 6.12).

As regards the yearly inflow of new migrants during the 1999–2012 lapse (see Fig. 6.13), it is possible to clearly distinguish two phases. The first, between 1998 and 2007, was characterized by the continuous and formidable growth of annual entries. With the exception of the year 2003, in which the increasing trend slowed down, in all the other years new records were registered. The maximum was reached in 2007, when a little more than 900,000 new migrants entered the country. The second phase, which started in 2008, was characterized by a decreasing trend. While the inflows remained sustained and exceeded 300,000 individuals per year, the change of sign was evident. In 2011, after 25 years of continuous growth, the immigrant population fell slightly, initiating a decreasing trend that persisted in the year to follow (Arango, Moya Malapeira, & Oliver Alonso, 2014).

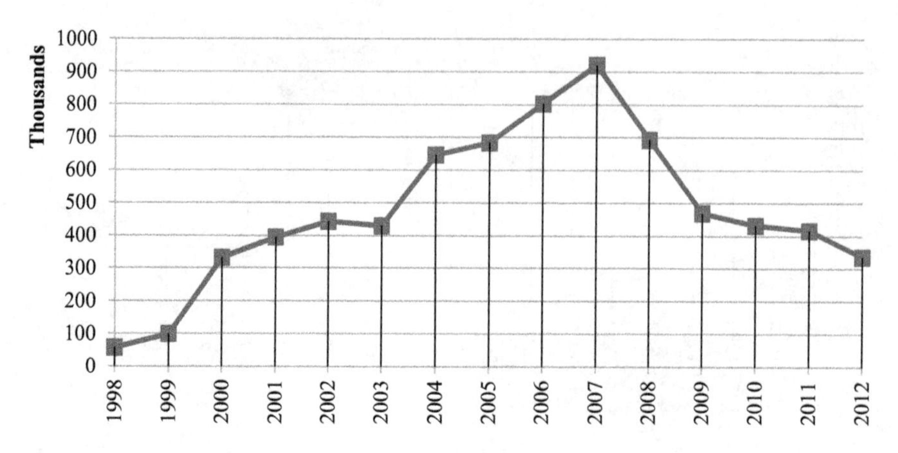

Fig. 6.13 Annual migration inflow. (Data from: OECD)

6.3.2 *Irregular Migration Estimations*

As pointed out by numerous scholars, the Spanish case provides a remarkably better possibility to elaborate irregular migration estimations than most of the other countries (Cachón, 2009; Cebolla & González Ferrer, 2008; González-Enríquez, 2009; Recaño & Domingo, 2005). This has been related to the strong incentive that irregular migrants have to register in the Municipal Records (*Padrón Muncipal*). This simple registration, which does not have any legal or administrative consequence, allows free access to most social services, such as, education for children or health care. The comparison between the total number of foreigners in the Municipal Register and that of foreigners with a valid resident permit (these include labour, study and asylum permits) allows one to obtain a fairly realistic estimation of the number of irregular migrants in Spain (G. Echeverría, 2010, 2014b).

As is possible to observe (see Fig. 6.14), also regarding the number of irregular migrants, two phases can be clearly distinguished. The first, between 2001 and 2005, was characterized by the sustained growth of the irregular population. The peak was reached in 2005, when estimated irregular migrants surpassed 1,400,000 individuals. The second phase, from that year on, displayed a sharp decrease in irregular population in the first 2 years, and stabilization with a decreasing tendency in the years to follow. The last available data, from year 2010, indicated a population of roughly 400,000 irregular migrants (Echeverría, 2014b). Two factors that contributed to the substantial reduction of the stock of irregular migrants registered in 2006 and 2007 were: A. the massive regularization enforced by the Spanish government in 2005; B. the automatic regularization of Rumanian and Bulgarian migrants determined by the admission of both their countries into the European Union on the first of January, 2007 (Finotelli & Arango, 2011; González-Enríquez, 2009).

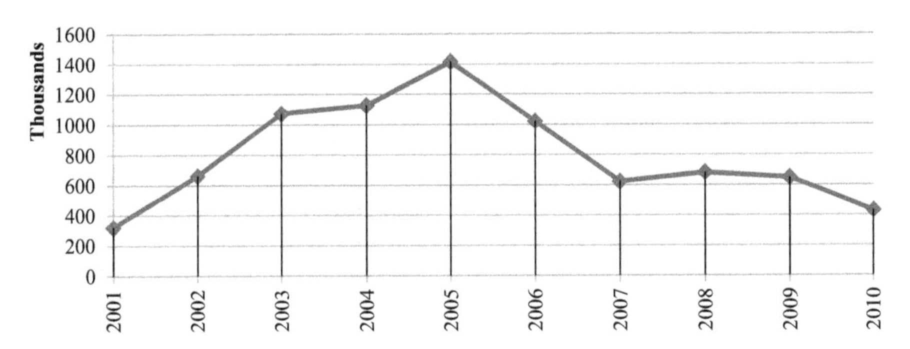

Fig. 6.14 Irregular migrants estimation. (Own elaboration, data from: Instituto Nacional de Estadística (INE) and Ministerio de Empleo y Seguridad Social (MESS)

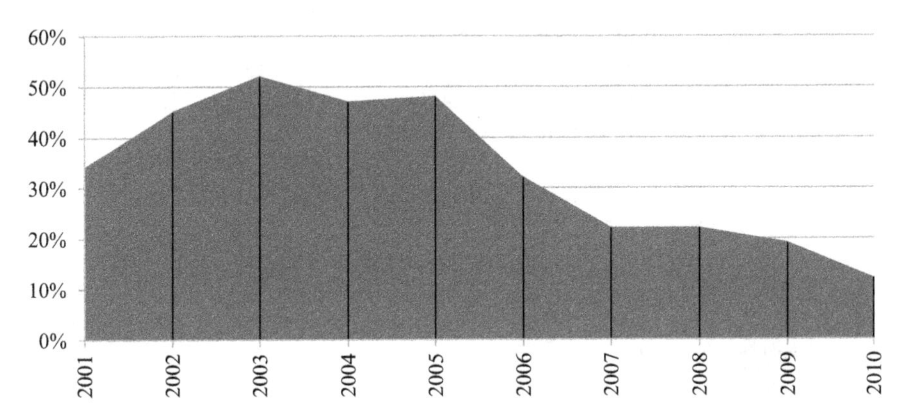

Fig. 6.15 Irregular migrant percentage over total foreign population (Own elaboration, data from: INE and MESS)

The share of the irregular population over the total foreign population (see Fig. 6.15) followed an increasing trend in the first years of the 2000s and reached its maximum in 2003, when it slightly exceeded 50%. In 2004 and 2005, the proportion decreased somewhat, but remained substantially above 40%. Also in this case, it is possible to clearly distinguish the combined effect of the 2005 and 2007 direct and indirect regularizations. In 2007, the irregular migration population accounted only for 20% of the total foreign population. The decreasing trend persisted in the years to follow. The last available data suggest that in 2010 the considered proportion had fallen to 12% (G. Echeverría, 2014b).

A number of studies have inquired into different aspects of the irregular migration population in Spain (Godenau, Hernández, & Expósito, 2007; Martínez Veiga, 2003; Recaño & Domingo, 2005; Van Nieuwenhuyze, 2009); a general overview of its main socio-demographic characteristics is available in Clandestino Report for the case of Spain (González-Enríquez, 2009).

6.3.3 Migration Regime

The Spanish migration regime is relatively short-lived (Aja, 2009b; Aja & Arango, 2006; Arango & Finotelli, 2009; Arango & Jachimowicz, 2005; Cachón, 2007, 2009; Cebolla & González Ferrer, 2008). This is not surprising if it is considered that until 1986 Spain had been one of the main emigration countries in Europe.

The first comprehensive migration regulations were approved in 1984 (Asylum and Refugee Law) and in 1985 (Foreigners Bill). Both laws had to be approved as part of the agreements contracted by Spain in order to become a member of the European Union. This circumstance had a fundamental impact on the regulatory conception that informed the two provisions. The main concern of the European partners, which in the majority of cases had a long migratory history and a restrictive attitude towards migration, was to avoid Spain becoming the new entry-gate for massive inflows. Moreover, the high unemployment rates registered in the country suggested that no labour migration was needed.

The 1985's Foreign Bill (*Ley Orgánica*[2] No. 7/1985) established a highly restrictive entry system for new migrants, the *Regímen General* (General Regime). The basic underlying principle was that before granting an entry permit to a migrant, a labour market check had to be carried out. Only if no native was available for the same position, could the migrant be hired and travel to Spain. In 1993, an additional mechanism was introduced, the *Contigente* (Entry Quotas). In this case, the administration, in agreement with the employer associations and the trade unions, had to establish each year a certain number of permits associated with available positions in the labour market to be offered to potential migrants. Neither of these channels ever worked properly. On the one hand, the *Regímen General* procedure was extremely complex and would have required a perfect coordination between the consular services abroad and the labour offices in Spain. On the other hand, also the *Contigente* required a complicated procedure and the pre-exiting agreement between the Spanish government and those of the potential migrants' countries. For this reason, as pointed out by Arango and Finotelli, this channel "never turned into an effective policy regulation instrument since it was simply used to legalize irregular migrants already living in Spain" (Arango & Finotelli, 2009, p. 18).

During the 1990s, as the Spanish economy started to grow consistently and the demand for foreign labour increased, it became apparent that a migration regime "imported" from countries with very different migration histories and labour market structures, was to be highly dysfunctional. Although the unemployment rate was high among natives, the segmented characteristics of the labour market determined the simultaneous existence of a high demand for unskilled foreign work. However, the available entry channels were insufficient and could not efficiently meet such demand. The combination of narrow channels for legal migration, embryonic migration control systems (since immigration was so recent) and an increasing demand for migrants exemplarily translated into an "irregular migration model" (Izquierdo,

[2] From here on, LO.

2009). Both for migrants and for employers, it was easier to achieve their respective goals independently of the channels enabled by the state.

As the alarm caused by the growing numbers of irregular migrants rose, a first major revision of the Foreigners Bill was approved in 2000 (LO No. 4/2000). The reform eliminated the *Regímen General*, reformed the *Contigente* and included the important decision to extend access to healthcare and basic education to all migrants without taking into consideration their administrative status. Despite these modifications, the entry regime remained largely ineffective and remained unable to satisfy the real necessities of the Spanish labour market (Arango & Finotelli, 2009). The problem was not solved either by the successive reforms approved thereafter, LO No. 8/2000 and LO No. 14/2003.

As had been occurring since the 1980s, the only effective measure to reduce the continuously growing stocks of irregular migrants was the implementation of massive regularizations. These "extraordinary measures" were "the most useful way to "repair" a posteriori, the structural mismatches of the Spanish migration regime in which irregularity and informality were constantly feeding each other" (Arango & Finotelli, 2009, p. 19). Between 1985 and 2005, the government approved six regularization processes (1985, 1986, 1991, 1996, 2000, 2001, 2005), which, altogether, rectified the administrative condition of 1,200,000 irregular migrants. The biggest regularization, called *Normalisación* (normalization), was ratified by the Socialist Party in 2005; this process alone involved more than half a million migrants. A prolific literature has analysed the characteristics, dimensions and consequences of these policy measures.

The 2005's regularization, however, was not just another episode of the well-known story. On this occasion, the measure was intended as part of a wide-ranging revision of the whole migratory regime that had started a year before with the approval of the Regulation 2393/2004. The new approach comprised four main lines of action, which have been thoroughly analysed (Arango & Finotelli, 2009; Cachón, 2009; Cebolla & González Ferrer, 2008; González-Enríquez, 2009). The first goal was to create adequate entry channels for foreign workers. In this respect, the *Regimén General* was re-introduced with a simplified procedure. A Catalogue of Hard-to-find-Occupations had to be published by the administration every 3 months in agreement with the trade unions and employer associations. An employer, who wished to hire a worker for a job that was included on the list, did not require a negative certification as had happened before. Moreover, modifications were introduced to *Contigente* and a new visa for "job search" was introduced.

The second goal was to create a permanent mechanism to allow irregular migrants to regularize on an individual basis. To this end, the *Arraigo* was introduced. This scheme permitted migrants to get a residence permit if they were able to demonstrate either a pre-existing labour story *(arraigo laboral)* in Spain or their social integration (*arraigo social*).

The third goal was to improve external border control in order to reduce irregular entries. A number of measures were taken, in particular: A. tougher rules as regards visa policy (in order to reduce visa overstayers); B. the introduction of sophisticated border control systems (in particular the *Sistema Integrado de Vigilancia Exterior,*

SIVE, in order to control the arrival of boats to the coasts); C. the development of bilateral agreements with the main sending countries (readmission and collaboration agreements).

The fourth goal was to reduce the attractiveness of the labour market and to make irregular residence more difficult by improving internal control policies. Two important measures were taken. On the one hand, the labour inspection agency was potentiated with more personnel and new strategies. This meant that the number of inspections per year was increased and their implementation was better targeted. On the other hand, a new emphasis was given to the repatriation policy. This implied more resources, newer and more efficient detention facilities, better identification systems, and agreements with origin countries.

A new reform of the Foreigners Bill was approved in 2009 (LO 2/2009) which was complemented by a new Regulation in 2011 (557/2011) (Aja, 2009a; Montilla, Rodríguez, & Lancha, 2011). These provisions extended the rights of irregular migrants in a number of sectors. In particular, the right to assemble, to associate, to demonstrate, to unionize and to strike was recognized. The possibility for irregular migrants to obtain free education was extended until they were 18 years of age. It was recognized that all foreigners, including those with an irregular status, had a right to have free legal protection in the case of need. However, at the same time, new restrictions were introduced. The family reunification policy was revised. As for irregular migration, a number of provisions were adopted to discourage irregular residence and employment and to make expulsions more effective; in particular: new infractions; higher fines for employers, traffickers, facilitators and migrants; new repatriation procedures; longer administrative detentions (from 40 to 60 days).

An important change that affected irregular migrants was introduced in 2012. The Real Decree 16/2012 excluded the possibility for those migrants without a valid residence permit to access healthcare assistance unless in cases of urgency, serious illness or accident (Montilla & Rodríguez, 2012).

6.3.4 Economics, Labour Market and Underground Economy

As can be observed in Fig. 6.16, the Spanish Gross Domestic Product (GDP), used here as a general indicator of the economic trends, shows two very different trends within the considered lapse of years. Between 1998 and 2008, on which the partial exception in 1999, 2003 and 2008, the economy markedly grew, registering positive variations that averaged 3% per year. Between 2009 and 2012, on the contrary, the economy underwent a deep recession. In 2009, the GDP variation registered −3.8%; in the next 2 years, it averaged a 0% variation; a new drop followed in 2012, with a − 1.6% variation.

The labour market followed a similar trend. Two contrasting, very marked phases are distinguishable. Between 1998 and 2007, there was a spectacular increase in total employment. In less than 10 years, more than 6,500,000 new jobs were created. In contrast, between 2009 and 2012, an accelerated destruction of jobs took

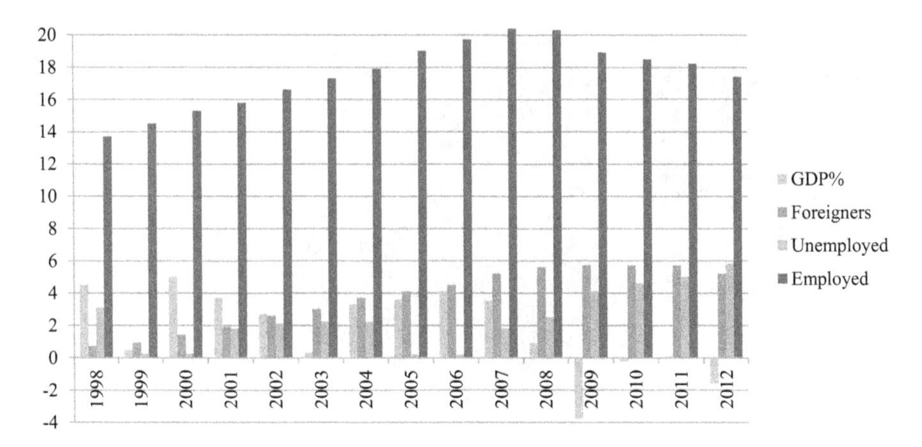

Fig. 6.16 GDP variation, employment, unemployment, migration. (Data from: Eurostat. GDP is expressed in annual variation; Foreigners, Employed and Unemployed are presented in millions per year)

place. The effects of the economic crises determined the loss of almost 3000,000 jobs in 5 year. Similarly to what was underlined in the discussion of the Dutch case, a direct, slightly delayed correlation with the GDP is observable. Yet, in the case of Spain, the effects of this relation appear to be much more accentuated (Finotelli & Echeverría, 2017). As pointed out by Finotelli, this has to do with the high level of elasticity of the Spanish labour market, which makes it very sensitive to the GDP variations (Finotelli, 2012, pp. 11–14).

As regards unemployment, an inverse, slightly delayed, very marked correlation with the GDP is observable. In general (see Fig. 6.17), if compared with its European partners, high levels of unemployment have characterized the Spanish labour market. In 1998, almost 20% of the active population was unemployed. The effects of the economic boom radically changed this picture in the next 10 years. In 2007, the unemployment rate had fallen to 8.7%. From that year on, however, the rate started to grow again, and progressively very quickly. In 2012, more than 24% of the active population was unemployed.

Concerning the occupation structure within the considered lapse of time (see Fig. 6.18), the Spanish labour market shows again two different phases. In the first, between 1998 and 2007, all occupations grew. However, the five sectors that created most new jobs were (International Standard Classification of Occupations - ISCO): Services and sales workers (+1.3 million), Technicians and associate professionals (+1.3 million) Elementary occupations (+1.1 million), Craft and related trades workers (+one million), Professionals (+1 million). In the next 5 years, while a total of almost 3000,000 jobs were lost, the distribution was uneven. The sectors where most jobs were lost were: Craft and related trade workers (−1.4 million) Elementary Occupation (−0.8 million), Technicians and associate professionals (−0.6 million), Plant and machine operators and assemblers (−0.6 million) and Managers (−0.6 million). The Services and sales sector (+0.7 million) and Professionals sector (+0.4

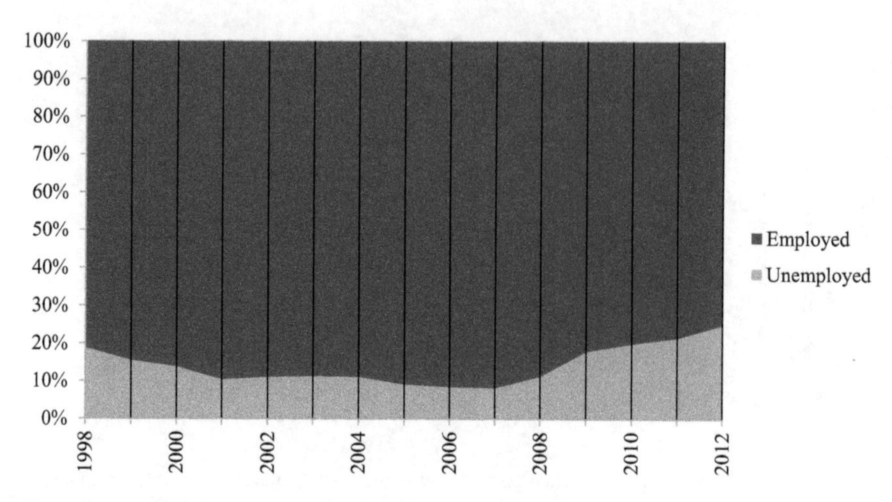

Fig. 6.17 Labour market structure. (Data from: Eurostat)

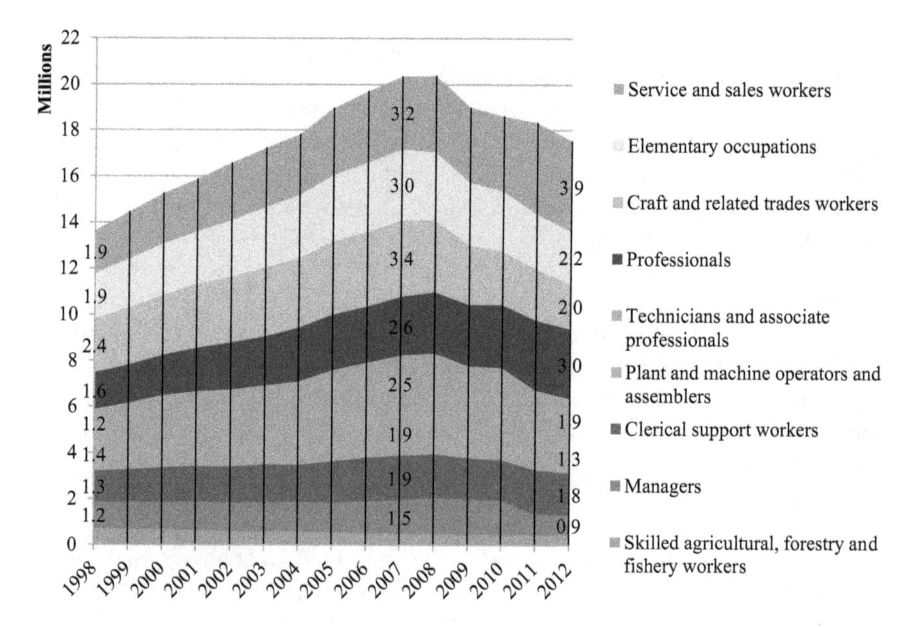

Fig. 6.18 Main occupations. (Data from: Eurostat)

million), on the contrary, continued to create new jobs. This analysis clearly shows the great importance played by unskilled sectors in the creation of jobs during the economic boom.

As regards the underground economy (see Fig. 6.19), the estimation provided by Schneider and his colleagues for Spain shows three different phases (Schneider et al., 2010, 2015). Between 1999 and 2003, the underground economy was stable, slightly above 22%. In the years to follow, until 2008, a decreasing trend was

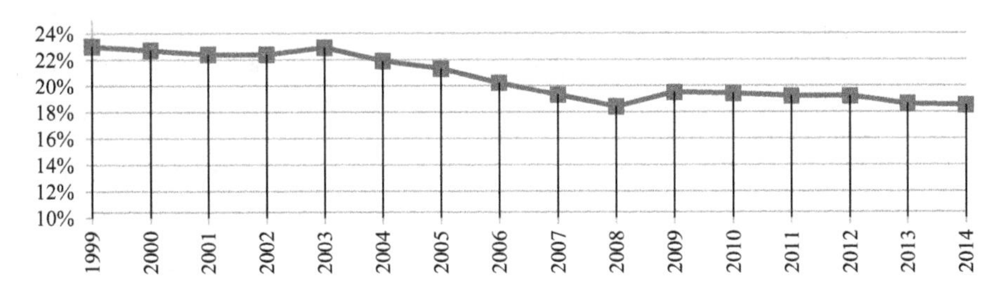

Fig. 6.19 Underground Economy. (Data from: Schneider et al., 2010, 2015)

observable that led to a fall of almost four points. From 2009 on, the rate has remained stable at around 19%.

6.3.5 The Welfare Regime in Spain

Within the different welfare state clusters in Europe, the Spanish welfare state is usually placed under the heading of the so-called Southern or Mediterranean Model (Ferrera, 1996; Ferrera et al., 2000; Gal, 2010; Hemerijck, 2012; Hemerijck et al., 2006, 2013). Besides the similarity with the Continental Model, the salient traits of the countries pertaining to the cluster, which also includes Italy, Portugal and Greece, model, are: the development of national health services; an acute insider/outsider distinction when it comes to social benefits; an emphasis on pension transfer in detriment to other social services; a stronger emphasis on the male breadwinner model combined with high levels of familiarism; weak or non-existent safety nets (Ferrera, 1996).

Beyond its intellectual interest, the long and on-going debate over the plausibility and usefulness of this fourth typology of welfare regime (Ferrera, 2008; Guillén, 2010; Guillén & León, 2011; L. Moreno, 2001), in addition to the three originally proposed by Esping-Andersen (Esping-Andersen, 1990), evidences two issues: on the one hand, the mixed nature of welfare regimes usually included within the Southern Model and, on the other, the continuous and deep transformations that these regimes have undergone in the last few decades.

These two issues perfectly apply to Spain. The Spanish welfare regime has been defined as a "hybrid of models" (Rodríguez Cabrero, 2011, p. 33). Its orientation appears Bismarckian, in relation to the income transfers and the emphasis on pensions, and Beveridgean, in relation to its universal national healthcare system. Moreover, the continuous procedure of reforms that the regime has undergone since its first development in the 1970s, makes it even more difficult to use a single label.

In its origins, the Spanish welfare regime was strongly influenced by the characteristics of its traditional society and the country's late modernization. This signified that it had a number of distinctive features (Hemerijck et al., 2013, pp. 33–34). First, there existed a pronounced insider/outsider cleavage between workers in the "core/

regular" sectors and workers in the "peripheral/irregular" sectors, the unemployed, family dependents or the poor. The former could rely on a generous system of social insurance, especially centred on pensions, while the latter were largely unprotected. Second, there was the paramount importance of families as the primary location of welfare production and economic redistribution (between generations and gender). In this context, the role of women was fundamental, determining a low level of female participation in the labour force. Third, there existed a highly regulated labour market, which fostered a marked dualism between permanent and temporary contracts. Fourth, social assistance programs were underdeveloped and weak which meant that there were comparatively higher levels of poverty (Rodríguez Cabrero, 2011, p. 33).

In the 1980s and the early 1990s, the Spanish welfare regime underwent a number of reforms that substantially modified its structure and scope. The Social-democratic model inspired the orientation of these interventions. The leading idea was that "subjective rights to health and education, financed through taxation" would "contribute to lessening inequalities and enhancing female access to the labour market" (Rodríguez Cabrero, 2011, p. 22). The main advancements were: the introduction of a universal education system (1985, 1990); the institution of a universal health care service (1985, 1990); the universalization of the pension system (1990); the introduction of regional minimum income Schemes (1989–1994).

Nevertheless, the economic crisis and the rapid rise in unemployment in the early 1990s, put the state budget under heavy pressure and forced the initiation of a process of welfare recalibration that, through a number of successive waves, has lasted until today. What has also contributed to this process was the concurrent Europeanization of social policy that implied the necessity to extend certain rights and to restrain expenditure.

In the important Toledo Pact (1995) "it was agreed that pensions and unemployment insurance benefits were to remain financed out of social contribution, but all the other non-contributory and social assistance benefits would come to be financed out of taxation" (Hemerijck et al., 2013, p. 35). The main lines of intervention in the years to follow have been four. A. Several measures were introduced to make the labour market more flexible and to balance the social protection between permanent and temporary workers. B. The social spending went through a process of rationalization and general reduction. A means-tested social assistance scheme (*Renta Activa de Inserción*) was implemented as well as the activation and formation of programs for the unemployed. Moreover, selective outsourcings and privatizations took place in the public welfare services. C. On the institutional level, the welfare services, including healthcare, education, care services were increasingly decentralized. D. Important measures were implemented to favour gender equality and the reconciliation of work and family life (Rodríguez Cabrero, 2011).

The effort to modernize and recalibrate the Spanish welfare state has been severely affected by the economic crisis that has affected the country since late 2007. The general budgetary cuts imposed by the economic situation signified a reduction of social expenditure, the termination of many social programs, a further flexibilization of the labour market, and a revision of the pension schemes (Hemerijck et al., 2013, pp. 34–37).

Three features characterize the contemporary Spanish welfare regime: first, the importance of the social security contributory system and the redistributive pension scheme; second, the existence of a universal system of education and healthcare not linked to labour participation; third, the still uneven and fragmentary development of the social assistance service. On the whole, then, as pointed out by Rodriguéz Cabrero, the Spanish welfare state has become "a consolidated medium-sized mixed welfare state with social spending levels below the EU-15 mean" (Rodríguez Cabrero, 2011, p. 25). Notwithstanding the important advancements in the last two decades, "it is the Bismarckian strand that still dominates the system as a whole; that is to say, what position in the labour market still counts more than citizenship, need or exclusion" (Rodríguez Cabrero, 2011, p. 34).

For the immigrant population, these characteristics of the Spanish welfare state have produced ambivalent results (J. Moreno & Bruquetas, 2012). On the one hand, universal access to education, healthcare and other social services for migrants, including those with an irregular status, has been exceptionally inclusive. Yet, access to healthcare was eliminated in 2012. On the other, the importance of the social security model and the fragmentary development of social assistance programs, have certainly be an element of weakness, especially considering the high levels of immigrant unemployment.

6.3.6 *Politics, Public Opinion, Migration*

Although definitely recent in the history of Spain, the migration phenomenon has strongly impacted its society. Statistics allow us to clearly measure the magnitude and speed of this change. In the lapse of two decades, the country passed from being a net emigration sender to being the second largest recipient of immigrants in the world, just behind the United States. In the decade of the 2000s, new arrivals reached extraordinary numbers. In 10 years the foreign population gained over five million individuals and their share of the total population grew from just under 4% to more than 14%.

While this spectacular transformation has certainly raised the attention of the public opinion and has materially changed the social landscape in many areas of the country, it has not led to significant anxiety or backlash. As pointed out by Arango: "Immigration was seen as a requirement of the labour market, an outcome of the economic progress, and perhaps even a sign of modernity" (Arango, 2013, p. 3). In his analysis, three arguments are proposed to support this claim. On the one hand, public opinion surveys have generally shown low, although slowly rising, levels of concern (Cea D'Ancona, 2011; Cea D'Ancona & Valles Martínez, 2013). There have been punctual moments in which the attention has risen, like during the *Cayucos crisis* (Cayucos are the small boats used by irregular migrants to reach the Spanish coasts) in 2006, but these have been rather exceptional. On the other hand, there has been no politicization of the issue. In Spain, until this day, no xenophobic or anti-immigration political party has obtained noteworthy consensus either at a national or at a regional level. The only exception, *Plataforma for Catalunya*, has not had any representatives at provincial, regional or national level. More in general, no party has used the anti-immigration discourse as part of its electoral strategy. Finally, and in connection to the previous point, "immigration policies have tended to be open, and integration efforts sustained and comprehensive" (Arango, 2013, p. 4). The efforts of the Spanish government, in contrast to what has been the general trend at a European level, have not included shutting down entry channels for migration. Instead, they have tried to improve the legal channels for immigrant workers and to establish permanent mechanisms for individual regularization (Arango, 2013, pp. 3–5).

As for integration policies, the Spanish government has shown strong commitment to immigrant integration (for a discussion of the different stands of the integration policy, see: Aja, Arango, & Oliver Alonso, 2012). Integration plans have been gradually developed at a national, regional and municipal level since the 1990s. Important consultative institutions, such as the Permanent Observatory for Immigration and the Forum for the Social Integration of Immigrants, have been created. In particular, the Forum, composed of nongovernmental organizations, immigrant associations, trade unions, employers' federations and the administration, has played a key role in orientating integration policies. The general orientation of integration policies has focused on the social and labour inclusion of the newcomers. Although there have been debates on the issue, until this day the Spanish approach has not followed the expanding trend to ask immigrants to pass language or civic knowledge tests.

Many observers expected that the positive attitude towards migration would have been negatively affected by the economic crisis. The impact of the economic crisis was indeed especially severe in Spain, and affected dramatically the immigrant population. Yet, as underlined by Arango, this circumstance "has not significantly altered social attitudes towards immigration, and immigration and integration policies have remained basically unchanged until now" (Arango, 2013, p. 6). Modifications to the migration regime (2009 and 2011), have not significantly altered liberal admission policies. Integration policies have been severely affected by the budgetary cuts introduced by the government at all levels; however, there have not been ideological reorientations or restrictive attitudes (Arango et al., 2014). An important exception to this generally preservative trend, was the approval in 2012 of a legislative decree which excluded irregular migrants, with certain exceptions (minors, pregnant women and emergency cases) from having the possibility to access public healthcare. Nevertheless, the application of this modification has encountered widespread social opposition and many regions have refused to operate it.

Arango has proposed three explanations for this generally positive attitude of the public and of the political world towards migration (Arango, 2013, pp. 9–12). On the one hand, the relative novelty of immigration to Spain and its high rate of labour participation have, for the moment, limited social conflicts. On the other, the peculiar historical and political evolution of Spain and, in particular, the recent regaining of democracy in the late 1970s, has contributed to generating a majoritarian political culture strongly influenced by democratic, egalitarian and universalist values. Finally, the absence of a militant national identity, motivated both by the multinational character of the country and the negative association of nationalism with the Franco regime, has inhibited ideas or feelings of immigration as a cultural threat.

6.4 Conclusion: Assessing Contextual Differences

6.4.1 Migration History and Contemporary Trends

Figure 6.20 shows an important difference between the Netherlands and Spain with regard to their immigration history. While the former is considered an old country of immigration, where second and third generations of migrants have grown up, the latter has a recent, although faster, migration history.

The Netherlands had a consistent immigrant population already in the 1960s, when it represented 4% of the total population. After a slight reduction in the 1970s, the share started to rapidly grow. In 1990, 8% of the population was born abroad. From that moment on, this share constantly rose, yet at a slower rate. In 2012, this was slightly below 12%.

In Spain, instead, the immigrant population remained under 2% until the 1990s. From that moment on, and especially after 1998, however, a spectacular increase took place. In 10 years, between 2000 and 2010, the immigrant population passed from 5% to above 14%.

Also regarding the recent immigration trends, the Netherlands and Spain display a very different picture (see Fig. 6.21). While fluxes to the former have maintained relatively stable averaging 90,000–100,000 new entries per year, the latter has experienced a “prodigious decade” (Arango, 2010, p. 54) of immigration.

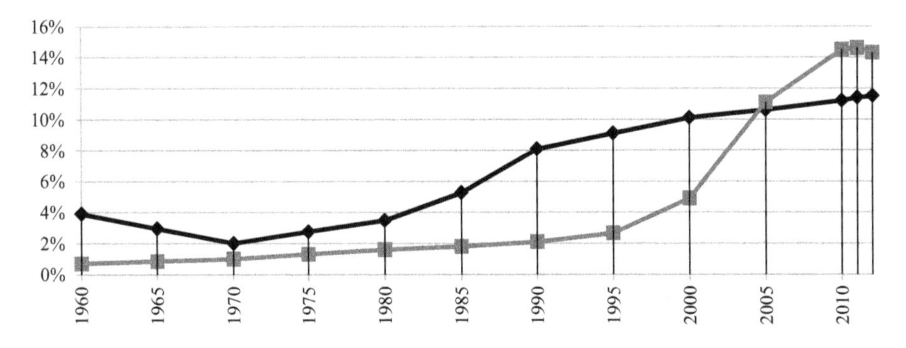

Fig. 6.20 Share of immigrant population over total population (Data from: OECD). Spain in Red, Netherlands in Light Blue

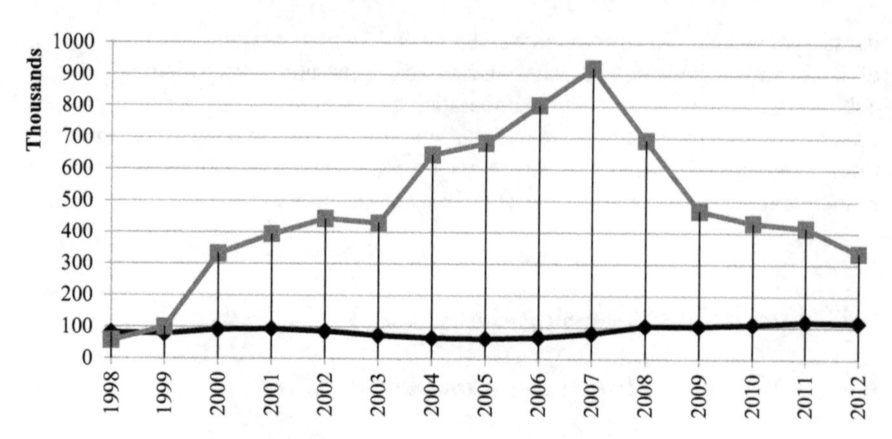

Fig. 6.21 Inflow of foreign population (1998–2012) (Data from: OECD). Spain in Red, Netherlands in Light Blue

6.4.2 *Irregular Migration Estimations and Trends*

The available estimations of irregular migration for the Netherlands and Spain in the decade of the 2000s display important differences regarding both their magnitude and trend (see Fig. 6.22). The maximum share of irregular migrants over the total foreign population was reached in the Netherlands in 2002 when it represented 30%. From that year on, a gradual but continuous reduction has taken place, and, in 2009, irregularity counted only for 15%. In Spain, it is possible to distinguish two different phases. Until 2005, irregular migration had a growing trend and was a huge phenomenon. The peak was touched in 2003 when the share rounded 50%, just like in the next 2 years. Between 2005 and 2007, there was a reduction of the stock of irregular migrants of almost 30 points, certainly the effect of the regularization of 2005 and the automatic regularization of Rumanians and Bulgarians in 2007. In years to follow, irregular migration in Spain has stabilized and appears to be slowly falling.

6.4.3 *Migration Regime*

In Table 6.1, a synoptic comparison of the actual migration regimes in the Netherlands and Spain is presented, with a specific focus on those elements that directly or indirectly affect the irregular migration phenomenon. Although in both countries legislation regarding migration has been continuously evolving, it is important to make a distinction as regards the extent of the changes in the period of our concern (1998–2013). While in the Netherlands, a number of modifications were introduced, it is possible to say that the basic normative model has been the same. In Spain, on the contrary, a crucial revision of the normative model took place

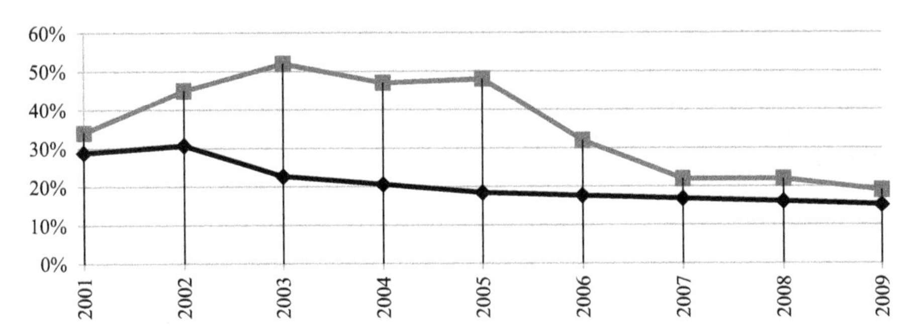

Fig. 6.22 Irregular migration share over total foreigners. (Own elaboration, data from: INDIAC – NL EMN NCP 2012, Eurostat, INE, MESS). Spain in Red, Netherlands in Light Blue

in 2004, so a clear distinction is possible between the period before and after that year.

As regards the actual migration regimes, at least as they appear on paper, and focusing on those aspects especially important in relation to irregular migration, three fundamental differences stand out between the Netherlands and Spain.

Firstly, there is an important difference concerning the available channels for legal entry. Considering labour migration for unskilled-workers, the Netherlands has a generally (there are limited exceptions) very strict, labour-check based, language and civic test limited admission policy. Spain has a flexible, labour-demand based admission policy. As for asylum policy, the Netherland has historically had generous, open policies with high degrees of demand acceptance (yet, this policy has become increasingly strict since the 2000s); Spain has historically had a very restrictive asylum policy with low numbers of demand acceptance.

Secondly, the Netherlands has had an exceptionally limited extraordinary regularization policy and has no permanent regularization schemes; Spain adopted recurrent, massive extraordinary regularization processes until 2005 and since 2004 it has had a permanent regularization scheme at an individual level.

Thirdly, the two countries have had very different approaches to internal migration control policies. In this respect, however, especially since 2005, Spain has been gradually moving in a direction closer to the Dutch one. In the Netherlands, since the late 1990s, there has been a comprehensive policy to dissuade irregular residence and work, and to enhance repatriations. The three pillars of this strategy have been: the exclusion of irregular migrants from social services and, in particular, from healthcare; tougher labour market controls (more inspections, assessing control responsibility to employers, higher fines); the improvement of identification technologies, detention facilities and re-admission agreements to improve expulsions.

In Spain, a wide-ranging policy to dissuade irregular residence and work through internal controls has been incrementally constructed only since 2004. The pillars of this strategy have been: the toughening of labour market controls and the improvement of expulsions policies. Contrary to the Dutch case, no exclusion policy was

Table 6.1 Synoptic comparison: migration regime and irregular migration

		Netherlands	Spain
Irregular migration status		Not a criminal offence	Not a criminal offence
Legal Entry Channels	**Labour migration**	a. Only if a Dutch or EU job seeker is not available.	a. Individual. If the job is included in a shortage list, an employer can directly make an offer and the job seeker can apply for a visa (Regimen General).
		b. Temporary work permit + temporary residence permits required before leaving home country.	b. Collective. Group recruitment, for specific jobs, from countries with a bilateral agreement (Contigente).
		c. Special schemes for large companies.	c. Job search visa. Visas are granted to job seekers for specific sectors.
		d. Special rules for highly-skilled migrants.	d. Special rules for highly-skilled migrants.
		Civic and language test required.	No civic and language tests required.
	Asylum seekers	Generous policy, high numbers.	Limited policy, low numbers.
Regularization Policy	**Extraordinary / Massive**	1975 (15,000);	1985–86 (23,000);
		1979 (1800);	1991 (110,000);
		1991 (2000);	1996 (22,000);
		1999 (1800);	2000 (152,207)
		2007 (27,500)	2000 re-examination (36,013);
			2001 (24,352);
			2001 (157,883);
			2005 (578,375).
		Total: 48,100	*Total: 1,103,830*
	Permanent / individual	Not available	Available (Arraigo)
Naturalization	**Through residence**	5 years of legal residence, proficiency in Dutch, knowledge of Dutch society (citizenship tests)	10 years of legal residence.
			2 years of legal residence for citizens of Latin American countries, Andorra, the Philippines, Equatorial Guinea or Portugal
	Through marriage	With a Dutch citizen.	With a Spanish citizen.
		With a EU-country citizen	With a EU-country citizen.

(continued)

Table 6.1 (continued)

		Netherlands	Spain
Internal Control Policies	**Access to social services for illegal migrants**	Education: free until 18 years of age.	Education: free until 18 years of age.
		Healthcare: free only for emergency cases.	Healthcare: free for irregular migrants until 2012.
			Other social services.
	Labour Inspection	Strict since 1998	Moderately strict since 2005.
	Expulsions	Increasingly since the 2000s	Increasingly effective since 2005.
	Random checks for documentation purposes	Not available.	Sporadic.

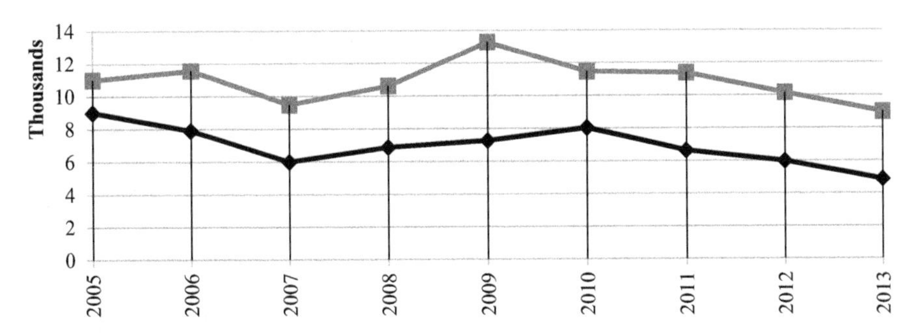

Fig. 6.23 Expulsions (Data from: NL (Data for Netherlands: 2005–2007, Leerkes and Broeders 2010, 2013; 2008–2013, Ministry of the Interior. Data for Spain: Ministerio de Interior). Spain in Red, Netherlands in Light Blue

enforced until 2012 and irregular migrants were able to freely access the public healthcare system and other social services.

Focusing on the efficacy of the expulsion policy, the available data (see Fig. 6.23), show similar trends between the two countries and moderately higher numbers for Spain. If the number of expulsions and the estimated irregular migrant population is considered, with the available data, a direct comparison is only possible for years 2005 and 2009. In 2005, the expulsion rate was 6.9% in the Netherlands, and 0.7% in Spain; in 2009, 7.4% and 2.0% respectively.

6.4.4 Economics, Labour Market and Underground Economy

As one can observe (see Fig. 6.24), the GDP of the two countries, between 1998 and 2013, shows different trends in the first years, until 2006, and a more similar picture in the years after that. In particular, the Spanish economy had an outstanding performance between 1997 and 2007 with yearly increases constantly above 3%. In the

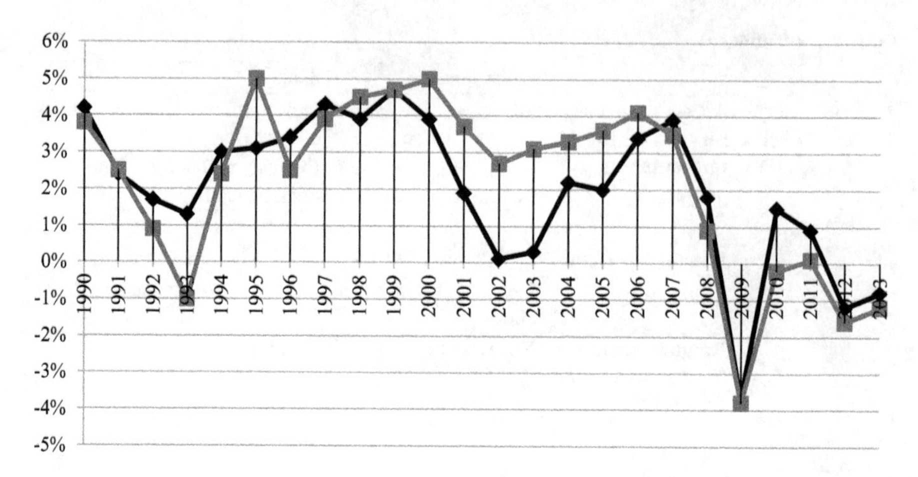

Fig. 6.24 GDP annual variation (Data from: OECD). Spain in Red, Netherlands in Light Blue

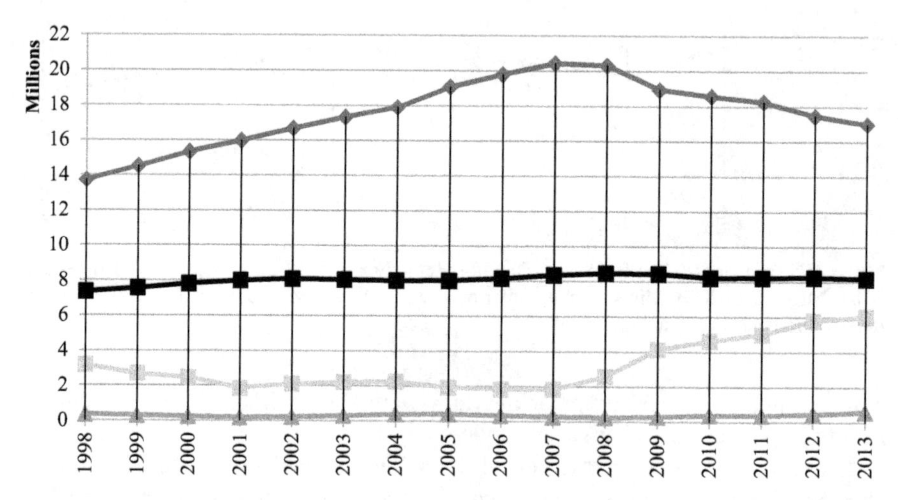

Fig. 6.25 Total employment and total unemployment (Data from: OECD) Spain in Red, Netherlands in Light Blue

same years, the Dutch economy had a much more ambivalent performance, especially in 2002 and 2003, when the GDP stagnated. The effects of the economic crisis struck the two economies severely in 2009. From that year on, the Netherlands had a slight recovery in 2010 and 2011, but the economy receded again in 2012 and 2013; the Spanish GDP, in contrast, never turned positive and was strongly affected by the new recession in 2012.

Considering the labour market (see Fig. 6.25), the pictures of the two countries are very different. The Netherland had a very stable tendency. Employment grew slightly, while unemployment had little variations. Spain, on the contrary, created more than 6.5 million new jobs between 1998 and 2008. Almost half of those, how-

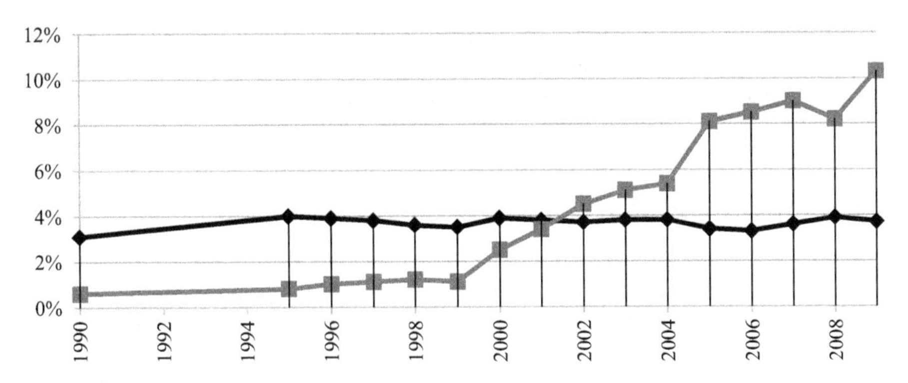

Fig. 6.26 Stocks of foreign-born labour (Data from: OECD). Spain in Red, Netherlands in Light Blue

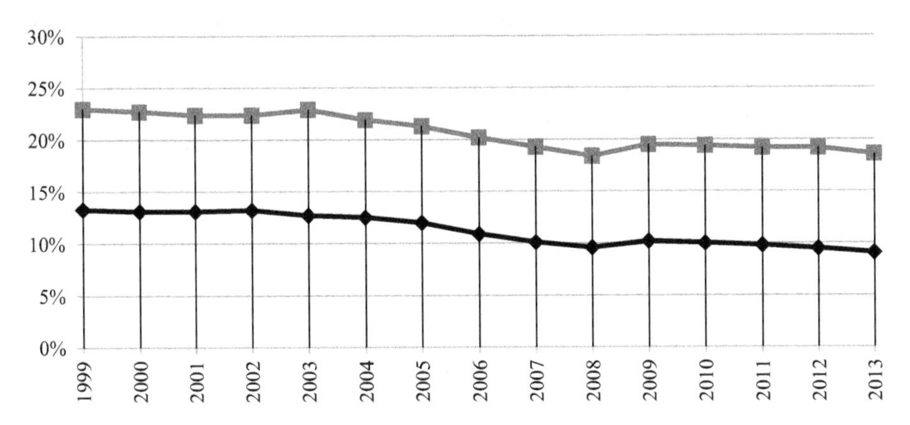

Fig. 6.27 Underground economy (Data from: (Schneider et al., 2010, 2015). Spain in Red, Netherlands in Light Blue

ever, were destroyed during the years of the economic crisis. Unemployment followed a similar trend, yet the effects of the crisis were even more marked. Between 2007 and 2013, almost four million individuals were registered on the unemployment lists.

In Fig. 6.26 it is possible to observe the significant relevance that foreigners played in the expansion of the Spanish labour market. While in the Netherland the share of foreign workers remained stable, in Spain between 1999 and 2009 it passed from around 1% to more than 10%.

The underground economy followed a similar slowly-decreasing trend in both countries (see Fig. 6.27). The size of the phenomenon, nevertheless, is significantly different. In Spain the shadow economy on average was 10% greater than in the Netherlands.

These data (GDP trends, employment and unemployment, foreigners in the labour market, shadow economy size) combined with those previously analysed on the different sectorial structure of the two labour markets (with a marked low-skilled orientation of the Spanish one) reasonably suggest that the Spanish economy has been much more attractive for irregular migrants that the Dutch one.

6.4.5 *Welfare Regime*

The Dutch and Spanish welfare states were both originally placed under the heading of the so-called Conservative welfare states (Esping-Andersen, 1990). It was in relation to the important differences existing between the northern and southern European countries pertaining to this cluster, of which the Netherlands and Spain are each paradigmatic examples, that Ferrera, in 1996, proposed the need of a fourth cluster of welfare states, the Southern or Mediterranean one. What the interesting and on-going scholarly debate about the pertinence of this new category indicates is that the existing differences are all but marginal.

Comparing the Dutch and Spanish welfare states, a first element of difference is purely quantitative. As one can observe (see Figs. 6.28 and 6.29), considering both the total expenditure per head and this as a percentage of the GDP, the Netherlands spent constantly and considerably more than Spain did, although there was a slow process of convergence.

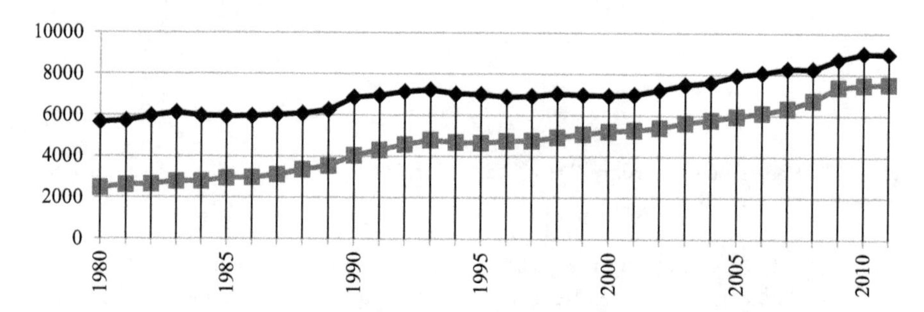

Fig. 6.28 Social expenditure per head at constant prices in US dollars (Data from: Eurostat). Spain in Red, Netherlands in Light Blue

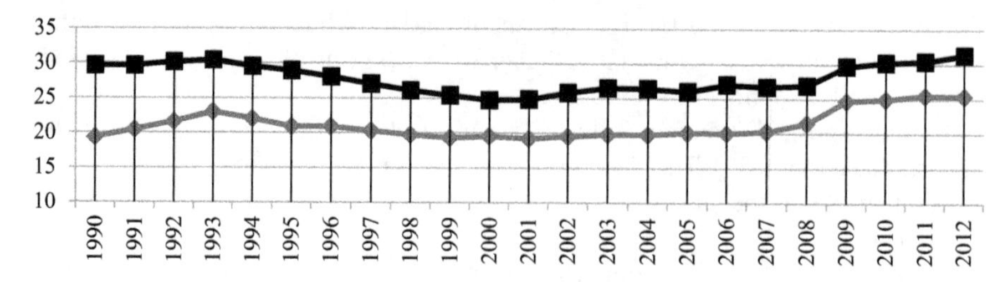

Fig. 6.29 Social expenditure as percentage of the GDP (Data from: Eurostat). Spain in Red, Netherlands in Light Blue

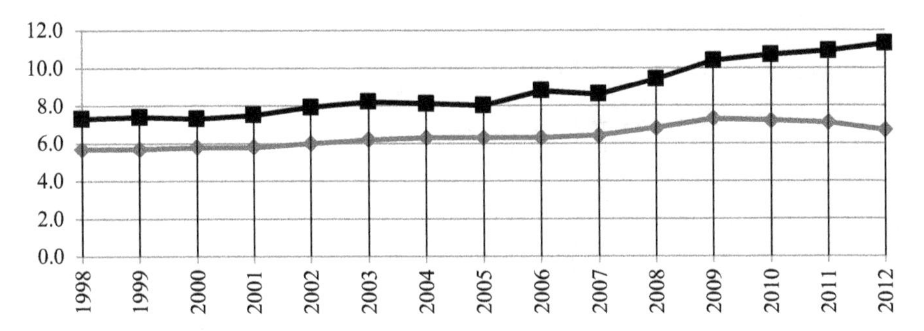

Fig. 6.30 Sickness and healthcare spending as percentage of the GDP (Data from: Eurostat). Spain in Red, Netherlands in Light Blue

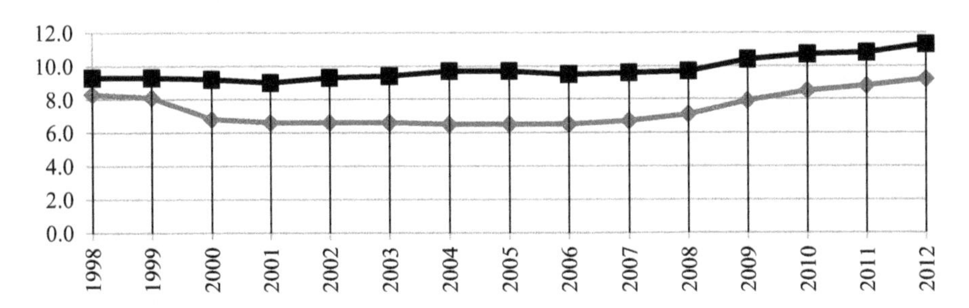

Fig. 6.31 Spending of old-age pensioners as percentage of the GDP (Data from: Eurostat). Spain in Red, Netherlands in Light Blue

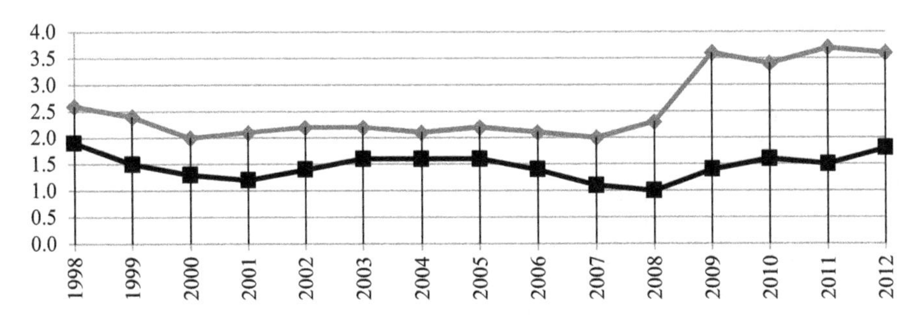

Fig. 6.32 Unemployment as percentage of the GDP (Data from: Eurostat). Spain in Red, Netherlands in Light Blue

In Figs. 6.30, 6.31, 6.32, and 6.33, it is possible to observe the social spending in the two countries in disaggregated terms. Each sector follows the general trend of the total figures. The only sector in which Spain spent more than the Netherlands is for the unemployment benefits. This is easily related to the different numerical relevance of the unemployed population. Noteworthy is also the case of Social Exclusion spending (Fig. 6.33) where a huge distance is observable between the two countries.

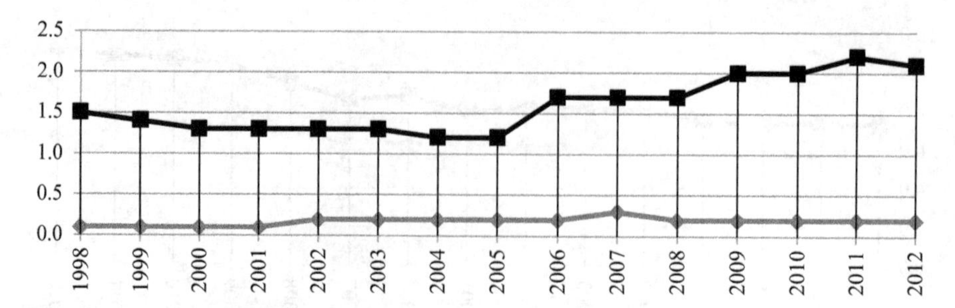

Fig. 6.33 Social exclusion spending as percentage of the GDP (Data from: Eurostat). Spain in Red, Netherlands in Light Blue

Hemerijck and his colleagues, more so on a qualitative level, have recently compared the welfare policies of the Netherlands and Spain (Hemerijck et al., 2013, pp. 8–9). In Table 6.2, it is possible to see the result of their comparison.

As can be observed, many relevant differences have been pinpointed. Whereas both welfare states have undergone processes of recalibration in recent decades, the Dutch government has been more effective in moving away from the limitations of the conservative model. In this respect, a number of important reforms have been introduced with the double objective of more efficient and universal social services, and a more flexible, yet supported, labour market (flexsecurity). In Spain, while important efforts have been made and noteworthy results have been attained, for instance, in the inclusion of women in the labour market, there is "the Bismarckian strand that still dominates the system as a whole; that is to say, what position in the labour market still counts more than citizenship, need or exclusion" (Rodríguez Cabrero, 2011, p. 34).

6.4.6 *Politics, Public Opinion, Migration*

Concerning the relation between politics, public opinion and migration, and given the complexity and sensitivity of the issue, it is certainly not easy to produce a concise comparison between the Netherlands and Spain. Therefore, it will be only possible to advance an impressionistic evaluation of the general trends based on the elements that emerged in the analysis previously proposed.

Following the analysis proposed by Arango, and using three elements (the public concern over migration, the politicization of immigration and the success of eventual populist parties, and the orientation of recent policy reforms and interventions in the immigration field) as a criterion to assess the general socio-political attitude towards migration, the Netherlands and Spain present a different picture.

Although the Netherlands had been a destination country beginning in the 1960s, migration became an issue of public and political concern in the 1990s. Since then and also in connection with a number of dramatic episodes at an international and national level, the social and political attitude towards migration has been increasingly complicated. The emergence of populist anti-immigration parties and their political success throughout the 2000s, whether they are interpreted as a response to

Table 6.2 Core principles of welfare regimes. (Hemerijck et al. 2013)

	Netherlands	Spain
Welfare regime type	Continental	Southern
Core values	Status preservation (equivalence principle)	Status preservation and differentiation
Objective	Income maintenance	Income maintenance
Social rights	Employment based entitlements	Insider biased entitlements.
Employment	Ambiguous work ethic (differences between Catholicism, Lutheranism and Calvinism)	Weak work ethic
	Full male employment	Full male employment
Gender	Nuclear family as cornerstone of society	Extended family as core welfare provider
Basis of entitlement	Work/family needs	Insider/family needs
Responsibility	Collective	Collective
Policy legacies, institutions and instruments of welfare regimes		
Social security	Social insurance financed high (contribution contingent) transfers (long duration)	Social insurance financed fragmented transfers (long duration)
	Separate public social assistance	No additional safety net
Labour market policy / regulation	Strong job protection, no active market labour policy	Strong job protection, no active labour market policy
Family support	Passive, but generous	Passive, but limited
Beneficiaries	Male breadwinners	Labour market insiders
Actors in provision	State secondary to the social partners (tripartims) and nuclear family (subsidiary)	Central role extended family (state rudimentary)
	Intermediary groups	Voluntary (church) organizations
Industrial relation	Sectorial-inclusive labour relation (wide coverage)	Politicized sector- and firm-based labour relations (fragmented coverage)

social anxiety or as its cause, certainly indicate a changed climate. Also considering the policies adopted since the 1990s, an increasingly restrictive attitude is evident. This has affected both migration control policies and integration policies. The former have been improved in a number of sectors. The objectives have been: to curtail entry channels both for legal and illegal migration, to discourage irregular residence and work, and to make expulsions an effective policy. As regards integration policies, both a discursive and practical departure from the multiculturalist paradigm has taken place. The new direction puts emphasis on civic and cultural integration and the acceptance of Dutch values as a necessary requirement for current and future immigrants.

The case of Spain has provided a different picture. It is certainly important to remember that the years in which migration emerged as a social problem in the Netherlands were the years in which it appeared as a social phenomenon in Spain. Yet, although conflictive episodes and moments of public concern over migration have existed, the general social climate towards migration can be judged as positive.

No populist, anti-immigration parties or discourses have emerged. This has translated into an open policy towards migration, which has centred on creating legal channels for labour migration and fostering the integration of the newcomers (Table 6.3).

Table 6.3 Synoptic comparison: the Netherlands and Spain

		Netherlands	Spain
Migration trends	**Historical**	Old country of migration (first, second and third generations)	Recent country of migration (first, forming second generation)
	1998–2013	Average 90,000 per year.	Average 476,000 per year.
	Irregularity	Moderate until 2002, low afterwards.	Very high until 2005, moderate until 2007, low afterwards.
	Ecuadorians	Very small community (3000)	Very big community (400,000)
Migration Regime	**Legal channels**	Narrow labour migration channels.	Narrow labour migration channels (until 2004); Flexible labour migration channels (from 2005).
		Broad asylum seeker channels.	Narrow asylum seeker channels.
	Regularization	Very sporadic and limited regularizations	Recurrent, massive regularizations
		No permanent regularization schemes.	Available permanent regularization schemes.
	Internal controls	Strict after 1998	Increasingly strict after 2005
		Irregular migrants excluded from healthcare and other social services since 1998.	Irregular migrants excluded from healthcare since 2012.
Economy, labour market, shadow economy	**GDP**	Booming economy 1994–2000 and 2006–2007 (GDP over 2.5%).	Booming economy1995–2007 (GDP over 2.5%)
		Mild economic crisis since 2009.	Deep economic crisis since 2009.
	Labour market	Slow growth of total employment.	Huge creation of jobs between 1998 and 2007 (+6.5 millions). Huge destruction of jobs between 2008 and 2013 (−3.4 millions)
		Unemployment: stable, very low unemployment	Unemployment: significantly decreasing until 2007; steeply rising in the years to follow.
	Sectors	Limited low-skilled sectors	Important low-skilled sectors.
	Shadow economy	14–9%	24–19%.

(continued)

Table 6.3 (continued)

		Netherlands	Spain
Welfare state	**Type**	Conservative	Southern/Mediterranean
	Main principles	Social insurance + Social assistance	Social insurance
	% of GDP	Between 25% and 30%	Between 20% and 25%
	Main universal services	Education, Healthcare, Old age pensions, Old age assistance	Education, Healthcare, Old age pensions
Politics, public opinion, migration	**Anti-immigration discourses and political parties**	Increasing importance of anti-immigration discourses in public and political debates.	No anti-immigrant discourses at a national level.
		Anti-immigration parties in the Parliament and in the Government.	No anti-immigrant parties.
	Public concern over migration	High to moderate.	Low to moderate.
	Restrictive integration policies	Increasingly restrictive policies since 1990s.	Liberal policies until 2012; then increasingly restrictive.

Bibliography

Acosta, A. (1998). *Breve historia económica del Ecuador*. Quito, Ecuador: Corporación Editora Nacional.

Aja, E. (2009a). La reforma de la Ley de Extranjería. In E. Aja, J. Arango, & J. Oliver Alonso (Eds.), *La inmigración en tiempos de crisis. Anuario de la inmigración en España 2009* (pp. 18–40). Barcelona, Spain: CIDOB.

Aja, E. (Ed.). (2009b). *Los derechos de los inmigrantes en España*. Valencia, Spain: Tirant lo Blanch.

Aja, E., & Arango, J. (2006). *Veinte años de inmigración en España: perspectivas jurídica y sociológica (1985–2004)*. Barcelona, Spain: CIDOB.

Aja, E., Arango, J., & Oliver Alonso, J. (2012). *La hora de la Intergación. Anuario de Inmigración en España 2011*. Barcelona, Spain: CIDOB.

Arango, J. (2005). La inmigración en España: demografía, sociología y economía. In R. del Águila (Ed.), *Inmigración: Un desafío para España* (pp. 247–273). Madrid, Spain: Editorial Pablo Iglesias.

Arango, J. (2010). Después del gran "boom": la inmigración en la bisagra del cambio. In E. Aja, J. Arango, & J. Oliver Alonso (Eds.), *La inmigración en tiempos de crisis. Anuario de la inmigración en España 2009* (pp. 52–73). Barcelona, Spain: CIDOB.

Arango, J. (2013). *Exceptional in Europe? Spain's experience with immigration and integration*. Washington DC: Migration Policy Institute.

Arango, J., & Finotelli, C. (2009). *Past and future challenges of a Southern European migration regime: The Spanish case*.

Arango, J., & Jachimowicz, M. (2005). Regularizing immigrants in Spain: A new approach. *Migration Information Source, 1*(9), 2005.

Arango, J., Moya Malapeira, D., & Oliver Alonso, J. (2014). 2013: ¿Un año de transición? In J. Arango, D. Moya Malapeira, & J. Oliver Alonso (Eds.), *Inmigración y Emigración: mitos y realidades. Anuario de la inmigración en España 2013* (pp. 12–27). Barcelona, Spain: CIDOB.

Boccagni, P. (2007). *Votare, per noi, era un giorno di festa. Un'indagine esplorativa sul transnazionalismo politico tra gli immigrati ecuadoriani in Italia.* Roma, CeSPI Working Paper, 35, 2007.

Broeders, D. (2009). *Breaking down anonymity digital surveillance on irregular migrants in Germany and the Netherlands.* Rotterdam, The Netherlands: Erasmus Universiteit.

Broeders, D., & Engbersen, G. (2007). The fight against illegal migration: Identification policies and immigrants' counterstrategies. *American Behavioral Scientist, 50*(12), 1592–1609. https://doi.org/10.1177/0002764207302470

Cachón, L. (2007). Diez notas sobre la inmigración en España 2006. *Vanguardia Dossier, 22*, 68–74.

Cachón, L. (2009). *La España inmigrante: marco discriminatorio, mercado de trabajo y políticas de integración.* Barcelona, Spain: Anthropos Editorial.

Cea D'Ancona, M. Á. (2011). Estabilidad y cambios de las actitudes ante la inmigración: un análisis cuantitativo. In E. Aja, J. Arango, & J. Oliver Alonso (Eds.), *Inmigración y crisis económica. Impactos actuales y perspectivas de futuro. Anuario de la inmigración en España. Edición 2010* (pp. 49–74). Barcelona, Spain: CIDOB.

Cea D'Ancona, M. Á., & Valles Martínez, M. (2013). *Evolución del racismo, la xenofobia y otras formas conexas de intolerancia en España. Informe 2013.* Madrid, Spain: Ministerio de Empleo y Seguridad Social – Observatorio Español del Racismo y la Xenofobia.

Cebolla, H., & González Ferrer, A. (2008). *La inmigración en España (2000–2007).* Madrid, Spain: Centro de Estudios Políticos y Constitucionales.

Cvajner, M., Echeverría, G., & Sciortino, G. (2018). What do we talk when we talk about migration regimes? The diverse theoretical roots of an increasingly popular concept. In A. Pott, C. Rass, & F. Wolff (Eds.), *Was ist ein Migrationsregime? What is a migration regime?* (pp. 65–80). Wiesbaden, Germany: Springer Fachmedien Wiesbaden. https://doi.org/10.1007/978-3-658-20532-4_3

Echeverría, J. (1997). *La democracia bloqueada. Teoría y crisis del sistema político ecuatoriano.* Quito, Ecuador: Letras.

Echeverría, G. (2010). *L'immigrazione irregolare in Spagna: fra politiche di controllo e crisi economica. Uno sguardo ai numeri. Neodemos.It.*

Echeverría, G. (2014a). Between territoriality, identity, and politics: The external vote of Ecuadorians in Madrid. In M. La Barbera (Ed.), *Identity and migration in Europe: Multidisciplinary perspectives* (pp. 175–191). Dordrecht, The Netherlands: Springer.

Echeverría, G. (2014b). De la "Producción institucional de la Irregularidad" a la "Irregularidad Sobrevenida": diez años de política migratorias en España. In M. Fernández (Ed.), *Negociaciones identitarias de la población migrante* (pp. 11–23). Madrid, Spain: Common Ground Publisher.

Engbersen, G. (2001). The unanticipated consequences of Panopticon Europe. In V. Guiraudon & C. Joppke (Eds.), *Controlling a new migration world* (pp. 222–246). London, New York: Routledge.

Engbersen, G., & Broeders, D. (2009). The state versus the Alien: Immigration control and strategies of irregular immigrants. *West European Politics, 32*(5), 867–885. https://doi.org/10.1080/01402380903064713

Engbersen, G., Staring, R., Van Der Leun, J., de Boom, J., van der Heijden, P., & Cruijff, M. (2002). *Illegale vreemdelingen in Nederland: omvang, overkomst, verblijf en uitzetting.*

Engbersen, G., & Van Der Leun, J. (2001). The social construction of illegality and criminality. *European Journal on Criminal Policy and Research, 9*(1), 51–70.

Engbersen, G., Van Der Leun, J., & Leerkes, A. (2004). *The Dutch migration regime and the rise in crime among illegal immigrants* (pp. 25–28). Presented at the fourth annual conference of the European Society of Criminology, Global Similarities, Local Differences. Amsterdam, August.

Entzinger, H. (2006). The parallel decline of multiculturalism and the welfare state in the Netherlands. In Banting & W. Kymlicka (Eds.), *Multiculturalism and the welfare state. Recognition and redistribution in contemporary democracies*. Oxford, UK: Oxford University Press.

Esping-Andersen, G. (1990). *The three worlds of welfare capitalism*. Princeton NJ: Princeton University Press.

Esping-Andersen, G. (1996a). After the golden age? Welfare state dilemmas in a global economy. In G. Esping-Andersen (Ed.), *Welfare states in transition: National adaptations in global economies* (pp. 1–31). London: Sage.

Esping-Andersen, G. (Ed.). (1996b). *Welfare states in transition: National adaptations in global economies*. London: Sage.

Ferrera, M. (1996). The "Southern model" of welfare in social Europe. *Journal of European Social Policy, 6*(1), 17–37.

Ferrera, M. (2008). The European welfare state: Golden achievements, silver prospects. *West European Politics, 31*(1–2), 82–107. https://doi.org/10.1080/01402380701833731

Ferrera, M., Hemerijck, A., & Rhodes, M. (2000). *The future of social Europe: Recasting work and welfare in the new economy*. Oeiras, Portugal: Celta Editora Oeiras.

Finotelli, C. (2012). *Labour migration governance in contemporary Europe. The case of Spain* (LAB-MIG-GOV Project – Which labour migration governance for a more dynamic and inclusive Europe). Torino: FIERI.

Finotelli, C., & Arango, J. (2011). Regularisation of unauthorised immigrants in Italy and Spain: Determinants and effects. *Documents d'anàlisi Geogràfica, 57*(3), 495–515.

Finotelli, C., & Echeverría, G. (2017). So close but yet so far? Labour migration governance in Italy and Spain. *International Migration, 55*, 39–51. https://doi.org/10.1111/imig.12362

Gal, J. (2010). Is there an extended family of Mediterranean welfare states? *Journal of European Social Policy, 20*(4), 283–300.

Garcés-Mascareñas, B. (2011). "Truble in Paradise". Reflexiones sobre los discursos y las políticas en torno al asesisanto the Theo van Gogh. In L. Cachón (Ed.), *Inmigración y conflictos en Europa* (pp. 231–270). Barcelona, Spain: Hacer.

Godenau, D., Hernández, V. M. Z., & Expósito, C. B. (2007). *La inmigración irregular en Tenerife*. Tenerife, Spain: Cabildo de Tenerife, Área de Desarrollo Económico.

Gómez Ciriano, E. J., Tornos Cubillo, A., & Colectivo, I. O. E. (2007). *Ecuatorianos en España: una aproximación sociológica*. Madrid, Spain: Ministerio de Trabajo y Asuntos Sociales.

González-Enríquez, C. (2009). *Undocumented migration: Counting the uncountable. Data and trends across Europe*. Country Report Spain. Report Prepared for the Research Project CLANDESTINO.

Guillén, A. M. (2010). Defrosting the Spanish welfare state: The weight of conservative components. In B. Palier (Ed.), *A long goodbye to Bismarck: The politics of welfare refroms in continental welfare states* (pp. 183–206). Amsterdam: Amsterdam University Press.

Guillén, A. M., & León, M. (2011). *The Spanish welfare state in European context*. Farnham, UK: Ashgate Publishing, Ltd..

Hemerijck, A. (2012). *Changing welfare states*. Oxford, UK: Oxford University Press.

Hemerijck, A., Keune, M., & Rhodes, M. (2006). European welfare states: Diversity, challenges and reforms. In P. M. Heywood, E. Jones, M. Rhodes, & U. Sedelmeier (Eds.), *Developments in European politics* (pp. 259–279). New York: Palgrave Macmillan.

Hemerijck, A., Palm, T., Entenmann, E., & Van Hooren, F. (2013). *Welfare states and the evolution of migrant incorporation regimes* (IMPACIM – The impact of restrictions and entitlement on the integration of family migrants). VU Univeristy of Amsterdam – COMPAS, University of Oxford.

Herrera, G. (2007). Ecuatorianos/as en Europa: de la vertiginosa salida a la contrucción de espacios transnacionales. In *Nuevas migraciones latinoamericanas a Europa. Balances y desafíos*. Quito, Ecuador: FLACSO.

Herrera, G. (Ed.). (2008). *Ecuador: la migración en cifras*. Quito, Ecuador: FLACSO-UNFPA.

Herrera, G., Moncayo, M. I., & Escobar, A. (2012). *Perfil Migratorio del Ecuador, 2011*. Geneva, Switzerland: Organización Internacional para las Migraciones (OIM).

Hoogteijling, E. (2002). *Raming van het aantal niet in de GBA geregistreerden*. Voorburg, The Netherlands: Centraal Bureau voor de Statistiek.

INDIAC – NL EMN NCP. (2012). In D. Diepenhorst (Ed.), *Practical measures for reducing irregular migration in the Netherlands*. Rijswijk, The Netherlands: INDIAC – NL EMN NCP.

INEC. (2013). *Anuario de Estadísticas de Entradas y Salidas Internacionales 2013*. Quito, Ecuador: Instituto Nacional de Estadística y Censos.

Izquierdo, A. (1996). *La inmigración inesperada*. Madrid, Spain: Editorial Trotta.

Izquierdo, A. (2009). El modelo de inmigración y los riesgos de exclusión. In *VI Informe sobre exclusión y desarrollo social en España: 2008* (pp. 599–679). Madrid, Spain: Fundación Fomento de Estudios Sociales y de Sociología Aplicada, FOESSA.

Kloosterman, R., Van Der Leun, J., & Rath, J. (1999). Mixed embeddedness: (In) formal economic activities and immigrant businesses in the Netherlands. *International Journal of Urban and Regional Research, 23*(2), 252–266.

Leerkes, A. (2009). *Illegal residence and public safety in the Netherlands*. Amsterdam: Amsterdam University Press.

Leerkes, A. (2016). Back to the poorhouse? Social protection and social control of unauthorised immigrants in the shadow of the welfare state. *Journal of European Social Policy, 26*(2), 140–154. https://doi.org/10.1177/0958928716637139

Leerkes, A., van San, M., Engbersen, G., Cruijff, M., & van der Heijden, P. (2004). *Wijken voor illegalen. Over ruimtelijke spreiding, huisvesting en leefbaarheid*. Den Haag, The Netherlands: Sdu Uitgevers bv.

Lucassen, L. (2001). A many-headed monster: The evolution of the passport system in the Netherlands and Germany in the long nineteenth century. In J. Caplan & J. Torpey (Eds.), *Documenting individual identity. The development of state practices in the modern world* (pp. 235–255). Princeton, NJ: Princeton University Press.

Martínez Veiga, U. (2003). Pobreza absoluta e inmigración irregular: la experiencia de los inmigrantes sin papeles en España. *Papeles de Economía Española*, (98), 214–224.

Montilla, J. A., & Rodríguez, J. L. (2012). Las normas generales del Estado en materia de inmigración en el año 2012. In *Inmigración y crisis: entre la continuidad y el cambio. Anuario de la inmigración en España*. Barcelona, Spain: CIDOB.

Montilla, J. A., Rodríguez, J. L., & Lancha, M. (2011). Las normas generales del Estado sobre inmigración en 2011. In E. Aja, J. Arango, & J. Oliver Alonso (Eds.), *La hora de la Intergación. Anuario de Inmigración en España 2011* (pp. 312–355). Barcelona, Spain: CIDOB.

Moreno, L. (2001). Spain, a via media of welfare development. In P. Taylor-Gooby (Ed.), *Welfare states under pressure* (pp. 100–122). London: Sage.

Moreno, J., & Bruquetas, M. (2012). Las políticas sociales y la integración de la población de origen inmigrante en España. In E. Aja, J. Arango, & J. Oliver Alonso (Eds.), *La hora de la Intergación. Anuario de Inmigración en España 2011* (pp. 158–186). Barcelona, Spain: CIDOB.

Penninx, R. (2006). After the Fortuyn and Van Gogh murders: Is the Dutch integration model in disarray? In S. Delorenzi (Ed.), *Going places. Neighbourhood, ethnicity and social mobility* (pp. 127–138). London: Institute for Public Policy Research.

Ramírez, F., & Ramírez, J. (2005). *La estampida migratoria ecuatoriana. Crisis, redes transnacionales y repertorio de acción migratoria*. Quito, Ecuador: Ciudad – Unesco – Abya Yala – Alisei.

Recaño, J., & Domingo, A. (2005). *Factores sociodemográficos y territoriales de la inmigración irregular en España* (pp. 18–23). Presented at the XXV International Population Conference, Tours, Francia.

Rodríguez Cabrero, G. (2011). The consolidation of the Spanish welfare state (1975–2010). In A. M. Guillén & M. León (Eds.), *The Spanish welfare state in European context. Farnham: Ashgate* (pp. 17–38). Farnham, UK: Ashgate Publishing, Ltd..

Schneider, F., Buehn, A., & Montenegro, C. E. (2010). New estimates for the shadow economies all over the world. *International Economic Journal, 24*(4), 443–461.
Schneider, F., Raczkowski, K., Mróz, B., & Futter, A. (2015). Shadow economy and tax evasion in the EU. *Journal of Money Laundering Control, 18*(1), 34–51.
Van der Heijden, P. G., Gils, G. V., Cruijff, M., & Hessen, D. (2006). *Een schatting van het aantal in Nederland verblijvende illegale vreemdelingen in 2005*. IOPS-Utrecht: Universiteit Utrecht.
Van der Heijden, P. G., Cruyff, M., & Van Gils, G. H. (2011). Schattingen illegaal in Nederland verblijvende vreemdelingen 2009.
Van Der Leun, J. (2003). *Looking for loopholes: Processes of incorporation of illegal immigrants in the Netherlands*. Amsterdam: Amsterdam University Press.
Van Der Leun, J. (2006). Excluding illegal migrants in The Netherlands: Between national policies and local implementation. *West European Politics, 29*(2), 310–326. https://doi.org/10.1080/01402380500512650
Van der Leun, J., & Bouter, H. (2015). Gimme shelter: Inclusion and exclusion of irregular immigrants in Dutch civil society. *Journal of Immigrant & Refugee Studies, 13*(2), 135–155. https://doi.org/10.1080/15562948.2015.1033507
Van Der Leun, J., & Ilies, M. (2008). *Undocumented migration, counting the uncountable. Data and trends across Europe. Country report The Netherlands*. 6th Framework Programme of the European Union.
Van Meeteren, M. (2010). *Life without papers: Aspirations, incorporation and transnational activities of irregular migrants in the Low Countries*. Rotterdam, The Netherlands: Erasmus Universiteit.
Van Meeteren, M., Van de Pol, S., Dekker, R., Engbersen, G., & Snel, E. (2013). Destination Netherlands. History of immigration and immigration policy in the Netherlands. In *Immigrants: Acculturation, socioeconomic challenges and cultural psychology*. New York: Nova Science Publishers, Inc.
Van Nieuwenhuyze, I. (2009). *Getting by in Europe's urban labour markets: Senegambian migrants' strategies for survival, documentation and mobility*. Amsterdam: Amsterdam University Press.

Chapter 7
Ecuadorian Irregular Migrants in Amsterdam and Madrid: The Lived Experience

In this chapter the results of the fieldwork realized in Amsterdam and Madrid will be presented. In particular, using the stories and information collected with the methodologies described in Chap. 5, a comparative analysis of the experience of Ecuadorian irregular migrants in the two cities will be discussed.

The objective of the fieldwork was to inquire into the main characteristics of the social experience of Ecuadorian irregular migrants in the two cities. In this respect, the collected material in each context was analysed and looked for possible regularities, behavioural patterns, and common experiences among migrants. Then, these results were comparatively analysed in an attempt to identify differences and similarities.

The analysis of the experience of irregular migrants was divided into two parts. Firstly, the main "legal trajectories" irregular migrants followed in the two contexts were identified. In particular, it was tried to find recurring patterns regarding both the length of the irregular status condition within the migratory trajectories and channels and strategies adopted by migrants to regularize. Secondly, it was tried to figure out what the living conditions of the migrants were when their status was irregular, and what the main related problems and the possible solutions were. The analysis focused on four areas: A. Regularization strategies; B. Work; C. Internal control experience; D. Housing and healthcare. In the conclusion, through a systematic and comparative analysis of the collected information, a general characterization of the irregular migration experience in the two cities was proposed.

Given the great amount of information collected, and its extreme richness, it was necessary to carefully select it. Inevitably, this process implied discarding interesting material and avoiding the discussion of many issues that emerged from the fieldwork. The selection was guided by two principles. On the one hand, it was privileged the material that was closely related to the phenomenon under inquiry. On the other hand, emphasis was put on those issues that could be compared more easily.

Although the qualitative nature of the data that will be presented in this chapter (and the methods used to collect them) were extensively discussed in Chap. 4, here

G. Echeverría, *Towards a Systemic Theory of Irregular Migration*, IMISCOE Research Series, https://doi.org/10.1007/978-3-030-40903-6_7

it is important to say a word of caution regarding their use. As it will appear, along the chapter, a number of hypotheses, inferences and extrapolations will be presented. It is important to keep in mind that, precisely because of the qualitative nature of the data used to support such propositions, these should not be considered as representative of the whole reality they refer to.

7.1 Legal Trajectories and Regularization Channels

In this section the attention will be devoted to the legal trajectories followed by Ecuadorian migrants in the two cities and to the available regularization channels. The concept of "legal trajectories" places the evolution of the migratory experience of migrants in relation to the administrative status allocated to them by the receiving states. The interviewed migrants were asked to recall their migratory stories, taking into consideration the evolution of their legal status and the main legal transitions they had gone through. The goal was to identify recurrent patterns concerning the length of the irregular condition, the channels used to regularize and the relevance of irregularity in their migratory trajectories.

7.1.1 Legal Trajectories and Regularization Channels in Amsterdam

Among the 30 Ecuadorian migrants interviewed in the city of Amsterdam, it was possible to identify 4 different legal trajectories (see Fig. 7.1). Two of these (A, B) were largely predominant; the other 2 represented minor, exceptional cases (C, D).

A. *Never Regular*

The first legal trajectory concerned the highest number of the interviewed migrants: 15 cases out of a total of 30. Although having resided, on average, 13 years in Amsterdam, the migrants of this group were not able to find any effective channel to regularize their status. In the majority of cases, they had implemented different strategies on numerous occasions to attain their goal, but their efforts were fruitless. Consequently, their administrative status has been irregular all along their migratory experience.

B. *Regularized through marriage or cohabitation agreements*

In numerical terms, this group counted 11 cases out of a total of 30. This trajectory sharply differs from the previous one; all migrants of this group were able to regularize at some point of their migratory experience. On average, the time needed to get a residence permit was approximately 8 years. The regularization

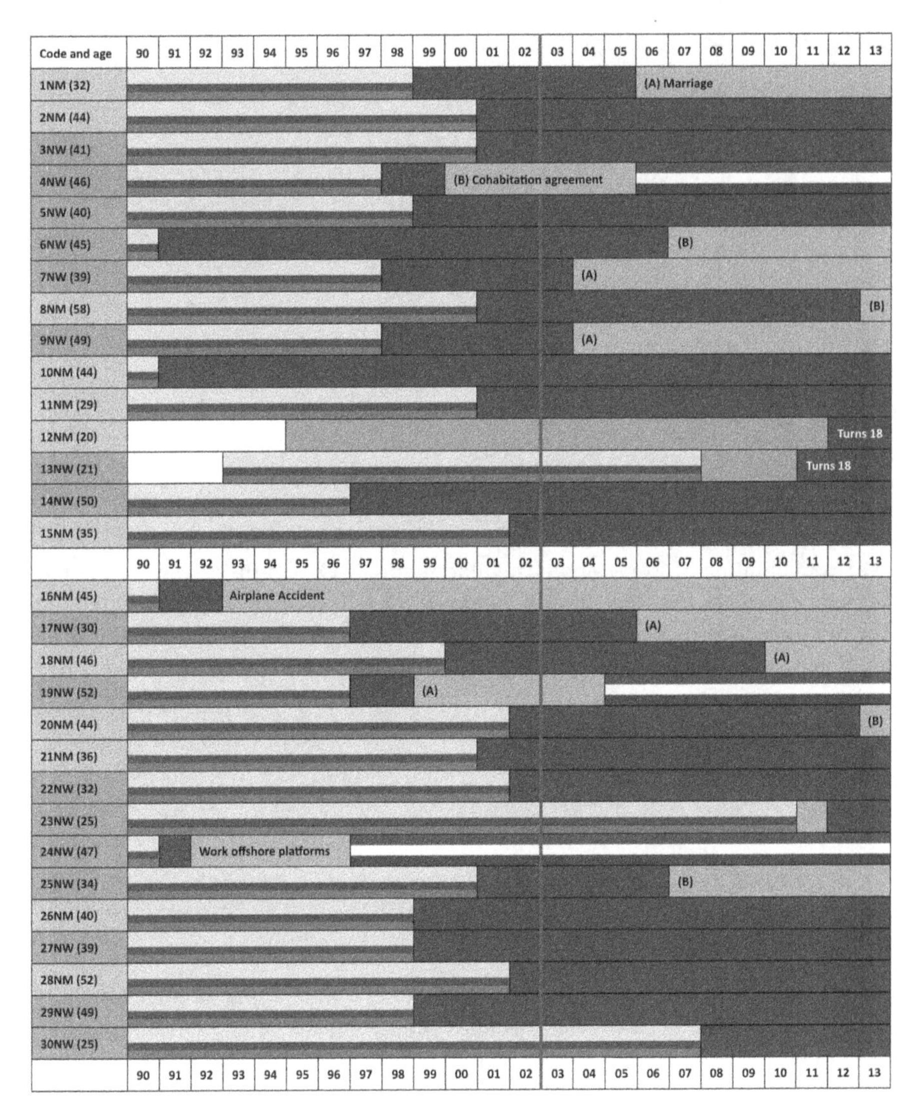

Fig. 7.1 Legal trajectories and regularization channels in Amsterdam (Own data)
Legend: on the left the code and gender of the interviewed (light blue – men, rose – women), in parenthesis the age; in yellow, blue and red bars the years in Ecuador; in red the years with irregular status in Amsterdam; in green the years with regular status in Amsterdam; in red, white and blues bars the years with Dutch citizenship. The red line indicated the introduction of visa request for Ecuadorians

channels used by this group of migrants were marriage (7) or cohabitation agreements (samenwonen) (4). In three cases, the migrants were eventually able to get a Dutch passport.

C. *Regularized under exceptional circumstances*

One migrant was able to regularize thanks to having legally worked for 8 years. After having overstayed her visa and lived irregularly for one year, in 1992, Marta[1] (24NW) was hired by an oil production company to work as a hostess in their off-shore platforms in the North Sea. The working contract offered her the possibility to eventually get a permanent residence permit after 8 years.

Another migrant was part of a group of almost 20 Ecuadorian indigenous musicians who obtained a permanent residence permit because they were involved in the "El Al Flight 1862" airplane accident. On 4 October 1992, a cargo aircraft crashed into a residential building in the Bijlmermeer neighbourhood. The great number of irregularly residing migrants involved in the accident, pushed the Dutch authorities to concede a residence permit on a humanitarian basis to all the migrants affected by the accident.

D. *Children of irregular migrants who become of age*

A slightly different trajectory involved two of the interviewed migrants. They were both children of Ecuadorian irregular migrants. One of them was born in the Netherlands. Since their parents did not have a residence permit, also their status was irregular. However, until their 18th birthday, while they did not have a residence permit, they could access public education and their lives were very similar to those of their schoolmates. The day after, they became "fully irregular", in the sense that they had to quit their studies and face all the difficulties connected with the lack of a permit.

7.1.2 Legal Trajectories and Regularization Channels in Madrid

Among the 30 Ecuadorian migrants interviewed in the city of Madrid, it was possible to identify 3 different legal trajectories (See Fig. 7.2). The first of these was largely predominant (A), the second represented a minor, yet relevant case (B), and the third was rather exceptional (C) (see Fig. 7.2).

A. *Regularized using legal channels*

The migrants who followed the first trajectory were the large majority of the sample; they were 21 out of 30. The common character of their trajectory was the effective and lasting regularization of their status using the *ad-hoc* legal channels. On average, the time needed to get a residence permit was 4 years. Regarding the type of channels, there were three available options: A. Extraordinary massive regu-

[1] For privacy reasons, all migrants' names that appear in the text are invented. The interviews' translation from Spanish is mine. For age and migratory history of each migrant check Figs. 7.1 and 7.2.

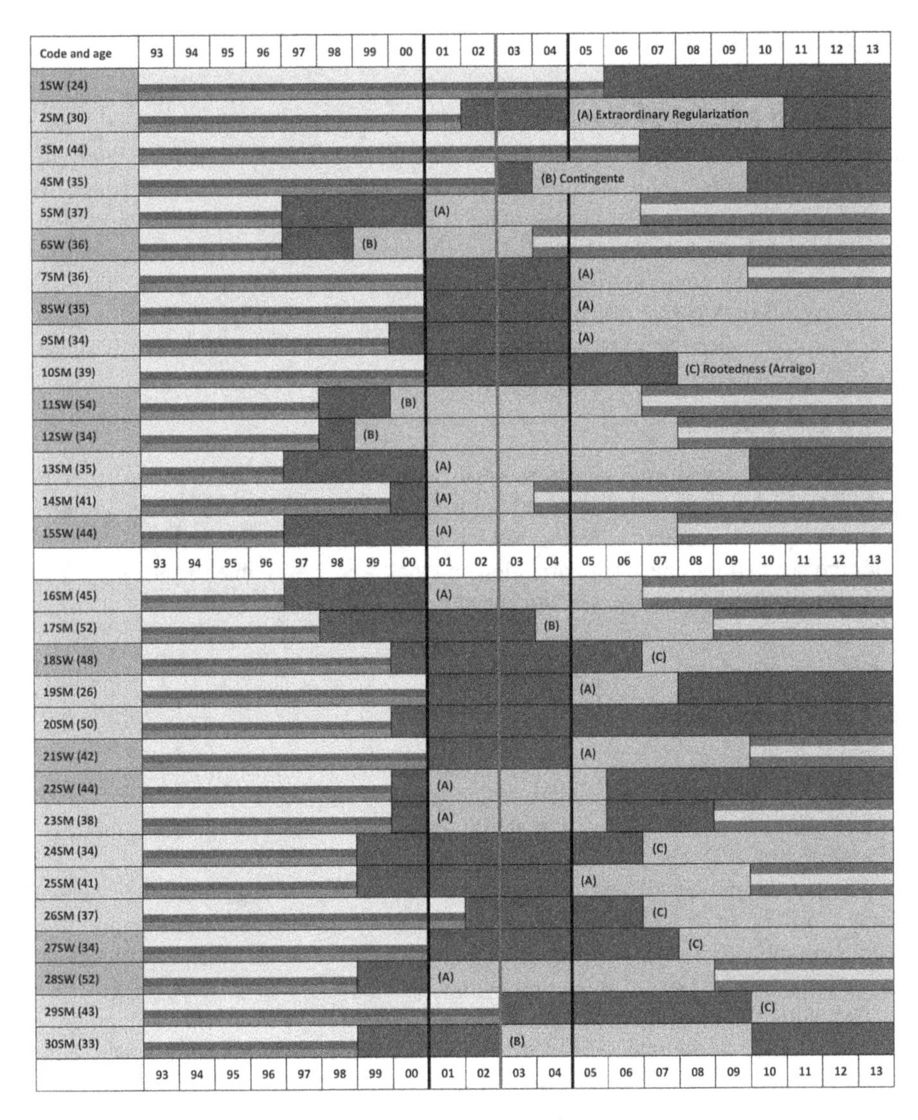

Fig. 7.2 Legal trajectories and regularization channels in Madrid (Own data)
Legend: on the left the code and gender of the interviewed (light blue – men, rose – women), in parenthesis the age; in yellow, blue and red bars the years in Ecuador; in dark red the years with irregular status in Madrid; in green the years with regular status in Madrid; in yellow and red bars the years with Spanish citizenship. The red line indicates the introduction of visa request for Ecuadorians. The black lines the regularization processes of 2000 and 2005

larizations (14 cases); B. Regularization through work quotas (*Contigente*) (5 cases); C. Regularization through rootedness (*Arraigo*) (5 cases). 13 migrants within this group obtained Spanish citizenship.

B. *Befallen irregularity*

The second trajectory, which concerned 6 migrants out of 30, was characterized by an initial phase of irregularity, a subsequent regularization through one of the available channels, and, finally, a return to irregularity. In all six cases, the return to the irregular status was determined by the impossibility of the migrants to renew their residence permit. The causes of such impossibility differed. In 4 cases, the migrants committed a crime after their regularization and when they had to renew their permit, they could not fulfil the clean police record requirement (the felonies were: assault on a public officer, driving under the influence (2) and domestic violence). In the other two cases, the migrants could not renew their permit because they did not have the required job offer.

C. *Never regular*

Three interviewed migrants were never able to regularize their administrative status. In one case, the migrant could not fulfil the clean police record requirement when he tried to regularize. In the other two cases, the migrants arrived in Spain after the last extraordinary regularization and they were never able to get a job offer that allowed them to use the *arraigo* channel.

7.1.3 Comparison

As can be observed, the legal trajectories followed by the interviewed Ecuadorian migrants in the cities of Amsterdam and Madrid display a number of significant differences.

Firstly, the number of migrants who were never able to regularize their status sharply differs. In Amsterdam more than one half of the interviewed migrants (17) were never able to regularize their status. In Madrid, in contrast, almost all the interviewed migrants (27) were able to regularize their status at some point.

Secondly, considering the length of the irregular phase within the whole migratory experience of the interviewed migrants, this averaged 12 years in Amsterdam and 5 in Madrid. For those who were able to regularize, the average time needed before getting papers was 8 years in Amsterdam and 4 in Madrid.

Thirdly, regarding the naturalization of former irregular migrants, in Amsterdam this was the case for 3 migrants, while in Madrid for 13 migrants.

Fourthly, excluding the exceptional cases, both in Amsterdam and Madrid, yet in very different proportions, it was possible to identify the *never regular* and the *regularization* trajectories. However, in the case of Madrid, a third relevant trajectory, not present in the Amsterdam case, appeared, i.e. the *befallen irregularity* trajectory.

Fifthly, fundamental differences emerged also in relation to the available regularization channels. In Amsterdam, excluding the two-mentioned exceptional cases, all migrants used the marriage or the cohabitation channels to regularize. It is interesting to point out, that, while these paths are perfectly legal, they were not, at least in the Dutch-state intentions, intended as regularization channels. Instead, in Madrid, three, *ad-hoc,* regularization channels were available. All the interviewed migrants consolidated their administrative status using one of these.

7.2 Regularization Strategies

Generally, regularization is not the first priority among migrants. For the vast majority of the interviewed, both in Amsterdam and Madrid, the initial idea when they travelled was to "make money" and go back to their country after a couple of years. The "issue of the papers" was either absent in their minds or openly considered irrelevant.

> I didn't know about the papers, I didn't care about the papers… The only thing I wanted was to pay back the money of the loan, to earn enough to build a house in Ecuador and go back…[2]

> A friend of mine told me about the papers… He said: I can help you to get the papers. I said: I do not want the papers, I have my job, I have my money, I will go back soon. What do I want the papers for?[3]

However, as the complexities of migration were revealed and it became apparent that return was not around the corner, all the migrants, although with different degrees of interest and determination, started to think about finding a way to regularize. Different reasons motivated this change of perspective.

> I started to think about the papers and then the visa for Ecuadorians was introduced. I felt as if I was in a cage. I could not travel or go back to Ecuador because, if I left the Netherlands, I was not going to be able to come back. Then, I said to myself: I have to get the papers at all costs!.[4]

> When I saw that the time in the Netherlands was stretching, I said: I need the papers to bring my children. I expected to go back in two years but after that, I had not yet obtained what I wanted. I could not stay any longer without my children.[5]

> I need to get papers because our children are growing up. Now they can go to school, but I heard that when they turn 18, they will not be able to go to school anymore. We don't want that; their education is the most important thing… and they cannot go to Ecuador, they are Dutch, they don't speak Spanish well… I am desperately trying to find a way, but here it is not easy…[6]

> I cannot go back without the papers… If you had children abroad… your children were not born in Ecuador, they were raised here, Sooner or later they will need to go out… If one day

[2] Interview with Marco (13SM).

[3] Interview with Elisabeth (4NW).

[4] Interview with Mauricio (18NM).

[5] Interview with Soledad (9NW).

[6] Interview with Lucía (14NW).

> I want to go back to Ecuador and the kids want to stay here, I don't want to force them to go. I want to prevent them from having to live what we had to. Going to a country where you do not belong, is tough. If I have a European passport I can go, I know I like Ecuador, it is my country! But if they don't like it, they can come back; they can study for career, you know, with a paper there is a guarantee.[7]

> For me the most important limitation of not having papers is that you cannot study, you cannot progress. I don't want to clean houses all my life… It is frustrating to know that you have no future, that your ambitions are blocked there… you want but you can't. Then, for me the priority has always been to get the papers.

Once the papers become a priority, the role of family members, friends or other migrants in providing the information about the options for regularization is fundamental. As a matter of fact, regularization channels, whether designed with that explicit purpose or not, in order to be effective, require a number of steps and fulfilments. Depending on the case, these can be relatively easy to accomplish or extremely difficult.

In this section, the focus will centre on the main strategies developed by migrants to overcome the difficulties in order to regularize their status in Amsterdam and Madrid.

7.2.1 Regularization Strategies in Amsterdam

As pointed out by Lucy (24NW) or Luis (28NM), in the Netherlands, it is very difficult for a migrant to regularize his/her status.

> The papers… that is more than a problem in the Netherlands… that is impossible here… it is very difficult… the only way is to get married, there is no other way. And also that is very hard, because they ask you hundreds of requirements. And now it is not even here, you have to go back and wait in your country. You have to pass an exam of the Dutch language that is very difficult. After all the years I have been here [22 years] I have not yet finished learning the language. Then, yes, to get papers is tough.[8]

> Here in the Netherlands you can do well… The only thing that you can't get are the papers… that no! It is a question of the state… of the law. They [the Dutch] don't know the word legalization… They did it, but a long time ago… those who benefited were the Muslims… a lot of them. But now they [the Muslim] don't behave well, they want their traditions and are rebelling… So, the Dutch said: No! No more! They stopped there… now the laws are very tough… Not because I have a Dutch passport, can I marry when I want and give papers to whoever I want… The moment I decide to do that I have to fulfil a lot of requirements…[9]

Excluding the exceptional cases, the only effective regularization channel has been through a recognized form of union (marriage or cohabitation agreements)

---

[7] Interview with Patricia (6NW).

[8] Interview with Lucy (24NW).

[9] Interview with Luis (28NM).

with a Dutch or a European-Union citizen. The two options implied a number of complex procedures and specific requirements. In both cases, the citizenship holder had to earn enough to be able to support the new companion and to have a house with adequate space for two; the migrant had to live in the same house as the companion; both had to be unmarried and the new relationship had to be considered plausible by the authorities.

According to the opinion of many migrants, the option to marry or sign a cohabitation agreement with a Dutch citizen has always been the most difficult one because the requirements were higher and strictly monitored. Moreover, it was necessary to go back to Ecuador and wait there for the approval of the process. Since the end of the 2000s, this option has become even harder because the migrants were required to pass a language exam before getting the residence permit.

> The only way to get papers here is doing *samenwonen* [cohabitation agreement]..., you cohabitate with someone and you get the papers... But if you do it with a Dutch person, they send you to Ecuador to learn the language. No! Now the key to get papers here, the Dutch papers, is to find an Italian, a Spanish person or an Ecuadorian or Colombian who has Dutch papers... I mean, that has a Dutch passport... then you get married... Not married, you do a *samenwonen*, as if you live together, they examine your case... and ta, ta, ta you show that you live together... you don't need to get married, nothing, and they don't send you to Ecuador...[10]

The best option, then, has always been to marry or sign a cohabitation agreement with a European citizen. The main advantage of this option was that it was not necessary to go back to Ecuador to wait for the visa. Another benefit was that the marriage could be contracted in another country, where the checks on salaries, houses, and veracity of the relationship were not so strict. Three of the interviewed migrants, for instance, travelled illegally to Spain to arrange the marriage there. Once the irregular migrant got the residence permit, it was enough to ask for recognition back in Amsterdam.

> It is difficult to find a Dutch citizen willing to arrange a bogus marriage. Those who accept are always in bad conditions... I mean, they are in drugs or have debts or are a bit crazy... Dutch people in their right mind would not do that... They are very serious. Those who can do that are Europeans. A lot of Spaniards, the majority, Italians, French. But the majority are Spaniards. The majority of those who got papers here was because of relationships with Spaniards. Perhaps it is also because of the language or because in Spain there are a lot of Ecuadorians and it is easier to get contacts. Moreover, if you make a deal with a Dutch person, it is much more complicated... You have to go back to Ecuador and that is a loss of time and money. Imagine. You have to go there, it can happen that you stay 2 years, and that means 2 years not working. When you get back, you have lost all your work, you are at zero. In contrast, if you arrange with a European, as long as the bureaucratic process goes on, you can work, you keep producing.[11]

For few migrants (2), this type of union arrived as the coronation of a love story; for the majority of the interviewed (9), it was merely a regularization strategy. In this second case, the unions were arranged on the basis of solidarity

[10] Interview with José (22NM).

[11] Interview with Maria (4NW).

(usually with friends or family members) (3) or, more often, as part of an economic transaction with the counterpart (6).

The development of a "marriage market" emerged as an adaptive solution to the narrow regularization channels and the high demand for them. On the one hand, the necessity for migrants to regularize at all costs, makes it reasonable for them to "invest" significant amounts of money in order to achieve their goal. On the other hand, for citizenship holders, the option to make some "easy money" without much risk could be attractive.

> Considering all costs, until now, I have spent almost 10,000 euro. Only to the girl I had to pay 8,000 euro, the rest has been for the lawyers… Yet, I am happy now, I think it was a good investment.[12]

> It is not easy but you find someone. There are many girls and kids available. They used to charge you 5,000 but now it is 8,000, 9,000 euro. Right now for less than 9,000 you don't find Dutch papers. If you have Dutch papers and you want to sell them to a girl, you know you have 10,000 euro…[13]

Yet, such transactions did not necessarily guarantee a good result. In many cases, the requirements were not met or the authorities suspected the veracity of the union. In those cases, new documents could be asked for or the permits could be simply denied. For the migrants, this could mean the beginning of a painful and costly *Via Crucis* of appeals and rejections that usually did not help them very much. Of course, once the deal was closed and the money paid, the outcome of the process was not a business of the vendor; for the migrants it was therefore impossible to get the money back or to retaliate somehow. In this type of transaction, the condition of legal weakness of the irregular migrants created the conditions for frauds.

7.2.2 *Regularization Strategies in Madrid*

In Madrid, the availability of a number of effective *ad-hoc* regularization channels, made it a lot easier for Ecuadorian migrants to normalize their status.

> For me it was very easy to regularize. After some three months working for a construction company, my boss said to me: do you want papers? I said: yes! I went to ask what I had to do at the foreign office. They said that what was needed was the registration to the municipal record and work contract and that my boss went there. I had already enrolled in the municipal register, because, everyone told you to register the day after arriving in Spain. So my boss went with the contract and everything was arranged. I had to go to Ecuador to pick up the visa, that was the only problem. But everything went well and after three months I already had my residence permit.[14]

[12] Interview with Manuel (20NM).

[13] Interview with José (22NM).

[14] Interview with Juan (4SM).

> With the extraordinary regularizations, it was super easy to get papers. You did not even need to go back to Ecuador. All you needed was the municipal record and a job… The only problem was to get the criminal record from Ecuador… I thought it was not going to arrive in time, my sister helped me in Quito…[15]

The three fundamental requirements, common to all the three previously mentioned legal schemes, were: the possession of a job offer, the presence in Spain before a certain date and the holding of a clean criminal record.

While these requirements were generally easy to fulfil, something that explains the high degree of regularization success, a number of strategies were developed in order to overcome possible problems. The requirement of a job offer in order to regularize and of a valid working contract in order to renew the residence permit, could be tricked with the use of false job offers or fake work contracts. In the latter case, the migrant would pay the social security and the contract costs to the employer so that he or she could pay them as if the migrant was effectively working. This type of strategy became particularly useful after the beginning of the economic crisis. The high levels of unemployment among migrants and the reduction of jobs more in general made it more difficult for migrants to satisfy the requests related to work. This affected both the regular migrants who had to renew their residence permit but no longer had a job, determining cases of befallen irregularity, and the irregular migrants who wanted to regularize using the rootedness channel (*arraigo*).

A distinction must be made regarding the criminal record issue. One of the main requirements needed to regularize through the extraordinary regularization processes was to present a document that certified that the migrant had not committed crimes in his origin country. If that was not the case, an option was to pay in Ecuador for a falsified criminal record. In the case of the renewal of a residence permit, instead, the requirement was not to have committed criminal offences in Spain. As seen in the situations of the 4 cases of befallen irregularity for criminal precedents, for migrants who could not fulfil this requirement, there was no strategy available.

> My residence permit expired three year ago… Everything went well, until the third renewal. I went there with all my documentations, I was relaxed, I didn't expect anything. Then they said that the permit was refused, that I had a problem because I had a criminal record. I was found driving under the influence of substances… I could not believe it… What happened was that once I was driving back home… I had been drinking a few beers with my friends and the Civil Guard stopped me. We are talking about 2 years ago… I had to do the alcohol test and I was above the limit. In that moment, I only had to pay a fine… but they did not tell me that it would affect the papers… but of course these things affect us… At that point I could not do anything.[16]

[15] Interview with Lorena (25SW).

[16] Interview with Ricardo (30SM).

7.2.3 Comparison

A fundamental difference is observable between the strategies developed in Amsterdam and in Madrid. In Amsterdam, where no *ad-hoc* channels are available, the majority of migrants who wanted to regularize their status had to find alternative ways. This situation triggered the elaboration of complex and risky strategies. Migrants' efforts focused on exploiting possible legal loopholes or finding ways to (mis-)use channels designed for other aims. The best example of this tendency was the use of marriage and cohabitation agreements as regularization channels. To this end, the strategies ranged from false unions with family members or friends, on a solidarity basis, to bogus unions with strangers, under payment. Another possibility was to travel illegally to countries where union requirements were softer and then to return to Amsterdam. A side effect, then, was the development of an underground business related to the papers and also the proliferations of frauds.

In Madrid, the vast majority of migrants could regularize without much effort thanks to the existing *ad-hoc* channels. Particular strategies had to be developed only by those who could not fulfil the requirements of those channels. This was, for instance, the case for the migrants without a job offer or a work contract, or for those who had a criminal record. In the first case, the solution was fake contracts, in the second, the falsification of the documentation. However, certainly noteworthy, the adoption of this type of strategies involved just a minority of migrants.

7.3 Work

In this section the focus will be on the work experience of Ecuadorian irregular migrants in Amsterdam and Madrid. Given the fact that many migrants were able to regularize, the attention will centre on their experience during the irregular phases of their migration trajectory.

Three main aspects will be analysed: A. the sectors where they were able to find work and the working conditions; B. the working conditions; C. the experience of controls on the worksites and the possible strategies to avoid them.

7.3.1 Work in Amsterdam

Sectors

Regarding the employment sectors of Ecuadorian irregular migrants in Amsterdam, a crucial element emerged from the fieldwork. If until the first years of the 2000s, both men and women had been able to find work in a relatively wide number of sectors, with the passing of time this number drastically fell to practically one sector.

In 2013, among the 15 migrants who still had an irregular status, 12 worked cleaning private houses, 2 worked in the construction sector and one worked as a domestic help in a private house taking care of children.

During the 1990s and the early 2000s, Ecuadorian irregular migrants were able to find opportunities in numerous sectors. Even though it is not too marked, a certain gender distinction was observable. The interviewed men had been employed in: hotels, cleaning rooms; restaurants, as dishwashers or cooking assistants; the construction sector, mainly as labourers and painters; the cleaning sector, both in offices and private houses; the port, as loaders; on the streets, playing musical instruments and selling handcraft products. Women had been employed in: hotels, cleaning rooms; restaurants, as dishwashers or cooking assistants; the cleaning sector, both in offices and private houses; as domestic help in private houses; on the streets, selling handcraft products. In many cases, the migrants had more than one job. A recurrent practice, especially among those who had a job with shifts, for instance, in restaurants or hotels, was to supplement their income by going to clean private houses in their free time.

From the early 2000s, the employment opportunities for irregular migrants started to fall. In many sectors, it became increasingly difficult to find work. Employers were no longer eager to employ migrants with an irregular status. For those migrants who had been working for a long time with the same employer, it was easier to keep their occupation. For those who had been fired and had to find a new employment the doors once open were now closed. This phenomenon was particularly evident in the service sector, for instance, in restaurants and hotels, and in the construction sector. By the end of the 2000s, almost no irregular migrant was employed outside the private house cleaning sector.

Asked about their opinion about the reasons for this change, the migrants offered two main explanations. The first was that, from a certain moment onwards, most employers stopped hiring irregular migrants. This was due to the increased fear of possible controls on the work sites and the risk of getting a fine.

> The work in the hotel slowly reduced. A lot of people were looking for jobs and the employers preferred to hire those with papers. Maybe in the high season they would still take you on, but only because they really needed workers . The city is full of people, a lot of tourists, so they need you. But now it is difficult because you have to have papers, even during the high season… Additionally, now that it is even harder, you are exploited… As time passes, the situation becomes more demanding, as everything does… Everything has changed… A lot! What can I say? I think that for more than 50% of us, almost 80%, 90%, things have changed. Even in restaurants now they don't hire you if you don't have papers. In hotels the situation is even worse. The market is dead, dead, dead… The only thing possible to survive now, because we still survive [the irregular migrants], is to work in houses… Why? Because they are private… it is private people that want you. There papers are not required, and there nobody stops you… The only thing that can stop you is if you don't have references, but nothing else…[17]

[17] Interview with Lucía (14NW).

> Now it is not as easy as before. Now nobody wants an illegal employee… In the past years many, many illegal worked here… In hotels, restaurants, in agriculture, picking tomatoes, but now nothing… Now you can only clean… clean, clean and clean… Nothing else. No company wants an illegal migrant…[18]

> Now the laws are very strict, nobody will expose himself/herself to hire a person without papers, nobody would risk the controls.[19]

> There was a lot of work… there is a lot of work! What happens, though, is that now the government is checking a lot… Before, they hired you "under the table", but now they don't. Now the situation is carefully controlled. Now you have to have papers, you have to have a working permit. So now the possibilities have reduced a lot, a lot! But, the possibility to work "under the table" is still available, for instance working in houses… It is not possible to work in restaurants and hotels anymore…[20]

> I had been working in that company for almost two years… We cleaned the windows of the big buildings… One day the inspections were made… I still thank the Lord because that day I was off… But the day after, I went to work, and my boss told me that I had to leave… that he was sorry, but I had to stop coming.[21]

> When I arrived here I could work in many places… they did not check… but that time has passed and that situation does not exist anymore… Today there is no employer that hires you if you have no papers. You need to be legal. The employer prefers to have legal workers, it does not matter if you are a migrant worker, but you have to be legal. The fines the employers get are very high… super high. So I stopped searching for jobs in other sectors. It was easier to work cleaning houses. It was also easier work because in the hotel I had to make 40 beds in the house 4… I quit in time, thankfully, because many flew to Ecuador because [she means deported] they were caught there working…[22]

Also those who worked in the streets playing musical instruments or selling handicrafts experienced the effects of a changed attitude by the authorities regarding informal work. They were unable to continue their activities without high risks.

> We worked playing musical instruments in the squares of the cities… We simply took out our musical instruments and played. We sold our cds and the handicrafts in the street… the police didn't say anything. Now the situation is fucked up! They stop you, they check you, they deport you. We have to run… Before we could travel around Europe, nobody asked us for our passports… Now it is impossible.[23]

The second reason given by the migrants was that new groups of migrants who had papers started to fill the labour market and take the jobs they used to get before.

> The first to come were people from the East… they have papers, they started to work where we worked before. Then a lot of migrants from Southern Europe started to come… They go

[18] Interview with Roberta (19NW).

[19] Interview with Johanna (27NW).

[20] Interview with Raquel (29NW).

[21] Interview with Lucho (1NM).

[22] Interview with Gabriela (27NW).

[23] Interview with Pablo (16NM).

where the economy is still good. Many Spanish Ecuadorians [Ecuadorians with Spanish nationality] are coming, because they have papers and find work…[24]

Now there are a lot of people coming… people that were in other countries. A lot of people from Spain, because of the situation there… But the situation for work became difficult when the European Union integrated… when those countries, those that are economically bad, where there is no work… So now, here there are more Polish than in Poland… They offer their work, they come to do it… There are Bulgarians… there are people from everywhere… So now, since a lot of people that come to work have papers, for those who are illegal it is more complicated… right now there are plenty of workers with papers.[25]

Yet, the shift towards the cleaning service in private houses was also part of a strategic option of the migrants themselves. Working in private houses offered a number of advantages in terms of salaries, flexibility of hours, security and work necessity.

When the controls started to become tougher, I was scared. I said: no! I won't look for jobs in hotels and restaurants anymore… In houses it is much better… There the people know you, they give you the keys, you go, you respect your schedule… the hours you have to work and you leave… You don't see anybody and it is impossible that they come to check you. Moreover, they pay you more… a lot more. In the hotels and restaurants they used to exploit me. So much, so much! Imagine, in the hotel Arena… They paid me 6 euro, in my houses I don't get less than 10… Can you imagine the difference?[26]

I had always worked in construction, at least for three years… But the problem with construction is that I had to work 10 hours per day, from Monday to Saturday. Of course it was good money… I used to get 600 euro per week, but it was only for short periods. I mean, all those jobs were temporary. It could happen that you stayed 3 or 4 months without a job… Then I got the first job cleaning in a private house. If you were able to find enough houses, it was much better… Much more stable… In construction it was always a problem, I worked for 2 or 3 months and then over… So I said: cleaning is much better, it is more stable and it is not so tough… You work inside… In construction you often have to work outside, in this ice-cold weather [it was February]…[27]

I used to work cleaning offices and cafeterias in the morning… But then I decided to quit and take more houses… Yes, because cleaning houses you earn the same or even more… you have to work less and have free time…[28]

Once we were playing music with a friend in the square… At one point the police arrived and asked for our passports… My friend had papers, so he started to talk to the policeman. He said a lot of things, that we were only musicians, that they should go in search of criminals… He distracted the police… I had the chance to run away… It was the second time in two months that I had to run away… So I talked to my wife, and we decided that it was better that I stopped working in the streets. The best thing was that I went with her to clean houses...[29]

[24] Interview with Xavier (23NM).

[25] Interview with Luis (28NM).

[26] Interview with Raquel (29NW).

[27] Interview with Mauricio (8NM).

[28] Interview with Patricia (4NW).

[29] Interview with Javier (23NM).

The advantage of cleaning houses is that it is impossible that they check you. You can go to the houses, clean and that's it. No, no, no… I never had problems, nobody ever has problems in the houses. Nobody has been caught working in a house and kicked out of the country. And many illegal people live even better cleaning houses… Yes, because it is safer…[30]

In that restaurant there were three illegal people working. My brother, a Turkish man and me. It was very hard… they really exploited you. The other workers with papers started to abuse… More work, more work and the same money. And you are not free to get ill, or to have a problem with your family… you have to be there always, 7 solid days there… They don't let you rest… and if you can't work they get angry. 'And you know what: I'll find someone else '. So at one point it was me who decided to quit. I had been helping sometimes my wife with the houses… I didn't like it because I thought it was a ladies' job, cleaning houses, ironing… I had some friends that did that job and earned their unfailing money… Sometimes I made fun of them… hahaha. I laughed… Now that is what I do! I have been doing it for more than five years… And now I have more work than my wife, I don't have enough time to get more houses. With this job I earn 2,000, 2,200 euro per month…[31]

The advantage of working in houses is that it is safer, quieter, easier, healthier… It is more relaxed, and the working hours are more flexible. If you have children, and one morning they wake up ill, you can call and say that you will go the day after… I mean, that possibility to suit your life… That for us who have children is crucial.[32]

Conditions

Notwithstanding the gradual reduction of sectors and increasing controls, Ecuadorian irregular migrants have generally judged their working conditions and opportunities in Amsterdam as good.

At the beginning it is hard, you don't know anyone… and it is hard. But then you start to know people, to make friends… They talk to you, they help you. Here there is a lot of work… once you start, you find more and more. There is plenty of work. And you can make money. Here the problems are others, the house, the papers, but there is work for everybody.[33]

Working in the hotel, me and my husband, we really made money. We had to work like mules… Maybe, now I think we were a bit exploited… Sometimes my husband started at 6 in the morning and finished at midnight… But we were able to save a lot and buy two houses in Ecuador…[34]

I think that the Netherlands gave us a lot… It has not been easy… you know… Because it is a very different country from ours… the language, the weather… the way the people are here. But we were able to work and send money back… Even though there is an economic crisis, I have a lot of work, sometimes I have to say no…[35]

---

[30] Interview with Gabriela (27NW).

[31] Interview with Pablo (28NM).

[32] Interview with Maria (4NW)

[33] Interview with Jorge (2NM).

[34] Interview with Maria (6NW).

[35] Interview with Pablo (28NW).

Many migrants, both regular and irregular, at the moment of the interview, agreed on the fact that, from the work perspective, the papers, paradoxically, may be a problem.

> I think that we who live here without papers live better… Those who have papers are all the time paying for something. Paying, paying, paying… In contrast, we who work under the table, let's say, we 'see' our money… Those who have documents, all their money goes away in payments. If we want to work, I mean, from Monday to Saturday, we can earn a lot more… What happens also is that those who get the papers kind of relax… They don't better themselves… they don't care anymore… It is also because when you are legal, the more you work, the more you pay to the social security. All our friends who have papers are always complaining that they have to pay too much…[36]

> When I finally got papers, the state helped me to find a job… It was in a storage centre… The pay was not bad, I don't remember, I think it was 1,200 euro. One day the boss asked me if I wanted to double the hours… I said: yes! I thought, if I earn 1,200 now, next month I would get 2,400 euro. The next month arrived I got 1,800 euro… I said: what?? You know… to work legally is a robbery… But the problem is that when you are legal they check everything… You cannot have more money… They want to know where you got it. For me, it was much better not to have papers…[37]

> For a long time we didn't even want the papers. You didn't need them. You had work, the kids could go to school and the people who had been able to get the papers said that they were in a better situation before… Because when they didn't have papers they had money and time, they could do everything… except traveling. With papers you don't have money, you don't have time and you cannot travel, because you don't have money and you don't have time…[38]

> Look, when you work here with papers, you get a basic salary that is enough for the basics. If you want to earn more here… you have to have some kind of qualification, you have to speak Dutch… Yet, if you work like us [he means irregularly] the more you work, the more you make. If you work from 8 to 8, let's say at 10 euro per hour, it is 120 per day… 5, 6, or even 7 days per week.[39]

> They say that when you are legal there is the advantage that if you lose your job you have unemployment benefit… But the truth is that if you have that… you are fucked!! They check everything about you. You have to study. You cannot miss a single day without a good reason. If you miss one day you are in troubles. You know, I am a musician. So if I go to work in the streets then I have coins. If you go with the coins to the bank they ask you: where did you get them? Did you work? If you spend a little more they check, if you spend a little less they check… In contrast, when I had my job under the table I could do whatever I wanted…[40]

[36] Interview with Luisa (29NW).

[37] Interview with Pablo (16NM).

[38] Interview with Mauricio (18NM).

[39] Interview with Lola (5NW).

[40] Interview with Pablo (16NM).

Controls and Strategies

Even if labour controls became increasingly severe through the 2000s, both migrants and employers in Amsterdam have always been alert to the possibility of inspections. For this reason, a number of strategies have been developed both to employ irregular migrants and to escape possible inspections.

Regarding the first aspect, e.g. the irregular employment of migrants, some sectors, for instance, port services, industrial cleaning and construction, appeared to have more inspections and required specific strategies. The decisive factor seemed to be the size of the business. When the employer was a medium-sized or big company, a contract was usually needed. The strategies, therefore, were basically aimed at bypassing this limitation. The two main options were: for migrants to rent or borrow the papers of a regular migrant; for employers, to hire more than one worker with a single contract.

> Once I worked in the port. We had to unload and load Russian ships… There you worked with the name of someone who had papers… Every morning when you arrived they told you: if the police of the port come, you have to say that this is your name… And you tired like hell, had to keep repeating to yourself who you were: Juan Charles, Juan Charles, Juan Charles… Sometimes they asked you just to check if you were alert. The "owner" of the job charged you a commission...[41]

> Sometimes I went to clean some big offices instead of my cousin… We were very similar, and her boss didn't say anything…[42]

To avoid controls migrants and employers develop specific strategies, which depend on the type of work. An important aspect, in all cases, is to try to pass unnoticed, especially when working on exposed sites.

> A lot of friends have been caught because they were working outside. You must always work inside because if you are working outside they can always ask you for your working permit.[43]

> I noticed that Carlos was very relaxed doing his job. He had to clean the stairs of a residence building. Yet, when he had to clean the street door of the building, the letterboxes that were outside and sweep the entrance of the building, he was nervous, worked at double speed and continually checked around. He said that that is the most dangerous part of his job… That when he is inside, nobody can check him.[44]

Many migrants agreed on the fact that in Amsterdam it is customary that inspections arrive because someone calls the police. Therefore, it is always advisable to be as little eye-catching as possible.

> Many times there were inspections when I was on the building site. I had to hide, go to the roof. They first enter and ask, if they don't see anything weird, nothing happens, but if they

---

[41] Interview with Juan (10NM).

[42] Interview with Luisa (29NW).

[43] Interview with Juan (10NM).

[44] Fieldwork note.

> see something suspicious, they call and more inspectors arrive. Many times, they come because there has been a complaint. Here [in the Netherlands] there are many complaints. A group of friends of mine, they were Brazilian, they were working on a building site in the street and they were listening to music that the people here do not listen to. Or examples of Ecuadorians listening to *salsa* or *bachata*… Then, they say: these are latino… For instance, when I work outside, on the street, for example painting, I always listen to a Dutch radio. If you want to listen to your music, use headphones and that's it. I always say to the new people, don't talk too loudly, don't sing… because here the people listen… I you are showy, to fail! But if you learn to be discreet, there's no problem.[45]

> You know, you have to be careful. When you work illegally in construction, sometimes the owner of another company or even workers may report on you. They don't like us, because we steal their work and often do the job for lower prices. Two friends of mine were deported because they had been painting a house. Suddenly, the police came… A cousin of mine was able to escape because he jumped down from the window. He said that he is sure that the guys working in the next house called the police…[46]

In restaurants and hotels, the owners always told the migrants where to hide in case of a labour inspection or gave them other instructions so as not to raise suspicions. In certain cases, they had a way to alert the workers back in the kitchen or in the corridors, about the arrival of inspectors, so that they had enough time to hide.

> In that hotel they told me not to wear the uniform, in case of a labour control I had to enter one room and pretend to be one of the guests. They also told me not to bring the bucket with the water or the trolley with the cleaning products into the rooms…[47]

> When I worked in that restaurant there were three inspections. They always turned on a light and I knew that I had to hide in the container of the dirty sheets… Once I had to stay there for 2 hours, I almost choked. I thought they had forgotten about me…[48]

> We knew what to do in case of inspections. In those years [before 2003] it was very unusual… but once we had an inspection. A Moroccan who was the oldest worker took me by the hand and we climbed from the stairs up to the roof. He told me to be careful because up there it was all greasy since it was where the extractors were released… After a while we heard something like a little bell, it was the cook beating with a knife on the metal… It meant the inspectors had gone… I didn't even see the guys of the inspection, their faces, what they looked like, how many they were. That was the only time, because before there were few inspections…[49]

---

[45] Interview with José (23NM).

[46] Interview with Pablo (16NM).

[47] Interview with Lola (5NW).

[48] Interview with Laura (24NW).

[49] Interview with Marta (7NW).

7.3.2 *Work in Madrid*

Sectors

Two important elements emerged from the fieldwork regarding the working sectors of the Ecuadorian irregular migrants in Madrid.

On the one hand, a drastic difference was noticeable between the years that preceded the economic crisis which started in 2008 and the years afterwards. The first phase was characterized by the abundance of opportunities in a variety of sectors; the second phase by a general and drastic reduction of such opportunities and the virtual disappearance of entire sectors.

On the other hand, the sectorial division between men and women, present also in the Dutch case, appeared significantly more marked in Madrid.

The combination of these two factors generated four slightly different labour-market and working opportunities for irregular migrants in Madrid: A. Men pre-crisis; B. Women pre-crisis; C. Men during the crisis; D. Women during the crisis.

Men in the pre-crisis phase were mainly employed in the construction sector. Among the 18 interviewed migrants, 14 had work at least temporarily in this sector. The other documented occupation sectors included: restaurants, industry, storage, transportation, and courier companies.

Women in the pre-crisis phase were employed in a variety of sectors with a certain prevalence of private-house cleaning and care work both for children and the elderly. Among the 12 interviewed migrants, 9 had worked mainly in private houses, 4 cleaning, 2 providing care to children and 4 providing care work for the elderly. The other documented occupation sectors included: restaurants, hotels, and professional cleaning.

While the effects of the crisis affected the whole labour market and, hence, also native workers and regular migrants, they were particularly tough for irregular migrants. As a matter of fact, the deterioration of the labour market conditions coincided with a restrictive turn on the part of the authorities. The combined effect was that the number of available positions fell and that for those positions many regular migrants were available.

The changed scenario especially affected men. The most important sector where they had found opportunities, e.g. the construction sector, literally collapsed.

> After one week that I had been here, a friend of mine took me with him to the building site. The boss said: perfect, you can start right away. After that, I always worked in construction. I had to adapt, to learn all the names, because in Ecuador we call the tools with other names… My boss helped me to get the papers… The first year without a contract I earned 900 euro, then when I got the papers I started earning 1200 euro. It was very good. One day, in 2009, the owner of the company came and said to us: that's it. There is no more work. He closed the company and that was the end… Now there is nothing… for 2 years I have been doing little things to survive.[50]

[50] Interview with Pablo (3SM).

I worked for more than 9 years in construction… At the beginning without papers and then with papers. I can tell you… That was crazy… we built, built, built… that seemed unstoppable… But we knew it could not last forever… we were building entire cities but there were no people… One day the company simply shut down and we were fired… From that moment on, it has been very difficult…[51]

For those who were still in an irregular administrative situation or who lost the papers, it became increasingly difficult to find opportunities to work. Among the interviewed, few had been able to keep their previous jobs in restaurants and in transportation; others started to find jobs in a sector that until that moment had been exclusively for women, e.g. house cleaning and care, while others relied on small occupations such as painting, gardening, electricity, etc.

I worked in that discotheque for more than five years… I had to clean and prepare everything for the next day… In 2007, the things started to go badly… Two of my colleagues were fired… My boss was very nice to me and he said that I could stay for some time. In 2008, they fired my boss and me… Luckily, I had unemployment benefit for more than one year… Now I basically have not worked for 3 years … I mean, sometimes a friend calls me for 1 month or little things… I am thinking of going back to Ecuador…[52]

Now with the economic crisis, for me it has become very difficult to find a job… Occasionally I find a couple of rooms to paint, a garden to look after or other small jobs like that [the term used is "*chapuza*" that literally means: 'work of little importance']. The reality is that now my wife is the one who is the bread winner...

When I could not renew the papers I didn't know what to do… There were no jobs for those with papers, imagine for me in that situation. I asked a friend of mine, who had been working for years with families, to help me. She introduced me to a woman she knew because she had worked at her house taking care of her father. They gave me an opportunity… I started working there… Thank god they didn't say anything about the papers… I think I was lucky, without my friend I would still be on the street.[53]

Right now my husband is in a worse situation than me… At least I have the chance to find hours or something [she means cleaning] in houses. He worked in construction for years, but now he has been unemployed for two years unemployed… Now it is a lot more difficult for men than for women.[54]

The situation was totally different… I know it because I was illegal at the beginning and I am illegal now. In the first years, I am talking about 2000, 2001, 2002, 2003… you just went to the square… There were queues of people waiting… all Ecuadorians… A car came and a man said: do you know how to pull cables? Yes! Jump in! It was super easy, then they knew you and they started calling every day… I worked in construction for years, I got papers, I earned money, I paid my debt back in Ecuador… When I could not renew my papers it was a cold shower… And so now I am illegal again. And now it is not like before… Imagine, my friends with papers are going back to Ecuador… Where do I find a job?[55]

---

[51] Interview with Fernando (26SM).

[52] Interview with Walter (16SM).

[53] Interview with Luis (30SM).

[54] Interview with Patricia (1SW).

[55] Interview with Xavier (13SM).

For women the situation has got worse as well. Yet, the cleaning and care sector seemed to be still offering opportunities.

> Until recently I was working with a cleaning company… The owner was helping me with the papers… I was there for two years. Two years ago, though, he said to me: 'you had better not work until you have papers'. The people working there were legal, but they hired me because my cousins told him about me. He agreed to hire me… But since things have become more difficult, he told me to stay at home. Now, I have been unemployed again for 4 months. It is difficult because everyone asks you for papers… For one hour or two that you want to work they ask you for papers…[56]

> Now it is has become very difficult. You have to have papers and you have to have references. For working with children or with the elderly, they ask you for references or the contact of someone that you have worked with. Before you simply went to a church and they gave you two or three telephone numbers of people looking for help at home. But now I have been to many churches to ask for work, since it is summer and there are a lot of people looking for someone who looks after the kids… But now everything is with papers...[57]

> Because now that the government said they were going to fine those who employ without papers, the people do not risk anymore. There is an association that helps migrants where they have a job board. You go and every job is with papers, with papers, with papers… I have been there and they say: whoever does not have papers, please leave, because all the offers are for people with papers… There is another association in Arturo Soria [an area of Madrid]. There is a girl there that helps people to find jobs, she is there every Tuesday and Wednesday. There is always a queue of women waiting. She comes out and says: 'Girls, there are opportunities for those with papers, those without must leave'…[58]

Conditions

In the years before the economic crisis, finding a job in Madrid was very easy. The vast majority of the interviewed said that it took them only a few weeks to start working; that there were many opportunities in different sectors; that nobody asked for papers. Regarding the working conditions and the salaries, the picture appeared more controversial. While in some sectors, such as construction, industry, transport, migrants usually had good salaries, relative stability and the possibility to regularize; in others, such as restaurants, private houses and stores, the situation was more variable.

> My boss was very nice, he really helped me a lot. I never felt exploited because I did not have papers. I knew how much those who were legal earned and it was the same as me… It was him who told me about the papers. His secretary did all the work, I had only to go and get them.[59]

---

[56] Interview with Patricia (1SM).

[57] Interview with Romina (8SW).

[58] Interview with Nuria (27SW).

[59] Interview with Daniel (7SM).

> I don't know how many people "made the papers" thanks to my company... They had many buildings... We worked a lot... I would not say it was easy... But you know, if you wanted, you could earn a lot, do extra hours, and everything. They helped us with the contract when there was the regularization...[60]

> There was a lot of work! A lot! I worked all day, but I made a lot of money. Imagine, I was able to send to Ecuador 800 euros per month. That in Ecuador was a fortune... I bought two houses, one for me and the other to rent...[61]

In general, when the labour relations were more "personal", the possibility for underpayment, exploitation or delays in the papers was more recurrent. In particular, the women who worked as domestic help in private houses were those who had a higher degree of bad experiences.

> In that house I worked from 7 in the morning to 10 at night when the kids went to bed... I had to do everything... Cleaning, cooking, ironing... everything. They gave me 450 euro... plus the room, but for all that work it was nothing...[62]

> What really upset me was that they did not help me with the papers. I did not know how to do them very well... And the woman said to me all the time: don't worry I will take care of them, I will take care of them... I stayed there for three years and they never did anything. All my friends were settling their situation... I don't' know why they didn't want to help me... maybe they didn't want to pay me more...[63]

> The owner of the restaurant was Ecuadorian... I can tell you... never work with the people of your country... They are the worst... I don't understand... Maybe it is because they come from the same place and now they feel superior... This guy made me work like crazy, every month he said to me: right now I only have this... Next week I'll give you the rest... bla, bla, bla... I left after 4 months... These people think that because you don't have papers you don't have dignity...[64]

Since the beginning of the economic crisis, the working conditions for irregular migrants have severely changed. On the one hand, the opportunities have fallen in every sector; on the other, the reduced opportunities have generated phenomena of downward competition.

> Until a few years ago, one used to work and send money to Ecuador. There were many who took advantage of their opportunities and made money. There are others who did not. But now, we are only surviving. Today there is no chance to really progress.[65]

> The problem is that the people are desperate to work. So, if you used to get 10 for one hour... now there are people who say 9, 8, 7![66]

[60] Interview with Victor (9SM).
[61] Interview with Alberto (23SM).
[62] Interview with Lucy (15SW).
[63] Interview with Marta (18SW).
[64] Interview with Hernán (24SM)
[65] Interview with Patricia (1SW).
[66] Interview with Luis (30SM).

Have you noticed how many flyers you find on car windows? How many flyers on the letter boxes? Painting, plumbing, gardening, removal, everything… These are people trying to do whatever they can. Every day, I go and post my little flyers… How many people call? You know, I think this is the right moment to leave…[67]

Controls and Strategies

Also regarding the labour controls and the strategies developed to avoid them, the situation in Madrid displayed two very different phases.

Until the mid-2000s before the start of the economic crisis, migrants' descriptions reveal a very relaxed situation. The vast majority of these migrants never experienced a control on a work site and the employers were not worried about hiring people with an irregular status.

"There was no problem… You know, everyone was illegal… so you just went and you started working. I think they knew that nobody was going to check, because otherwise they would have been more worried…".[68]

In the construction? Never… never a single control…[69]

In the restaurant where I worked for more than 8 years, we never had a control…[70]

In all sectors there was work under the table…, in all sectors: painting, plumbing, construction, transport… And they could not check… I think it is too difficult. I don't know if they don't want to or they can't. For instance, I have always worked in the transport and removal sectors. Until recently, there were no checks at all… They should give fines to those you contract the service with… But they try to catch whoever is doing the work… and that is very difficult. How can you prove that they are working, that it is not a private thing...[71]

Working in houses, the control is impossible… It is the safest work… I never heard of anyone being checked or anything. Even now that they say controls are tougher, it is impossible. It is more dangerous because when you work in houses you have to move around the city, with the metro, with buses… That could be dangerous.[72]

From the mid-2000s on, and especially in certain sectors, a gradual increase of controls was recorded. The employers, who until that moment had been basically unconcerned, started to ask more frequently for papers or to develop strategies to avoid possible controls. Accordingly, also irregular migrants had to develop their own strategies in order to get hired.

[67] Interview with Fernando (26SM).

[68] Interview with Jesus (20SM).

[69] Interview with Alberto (23SM).

[70] Interview with Marcelo (6SM).

[71] Interview with Carlos (24SM).

[72] Interview with Lorena (27SW).

> They kept hiring irregular migrants. It became only a little more difficult. When I finally regularized… My name is Xavier Ramirez, we went to the working site and there were three Xavier Ramirez… My boss said to me: don't work there… I asked: why? He said: because there are two others with your name. I said: but you pay me the day? Yes! Since he had my documentation he could do that… I went back home but he had to pay me the day. Why? Because he had my papers. Before, I could not say anything because it was me who was the one working with the name of another… But now…

> When controls increased, a lot of people made money acting as intermediaries. There were Ecuadorians and Peruvians who really made lots of money. They had contacts with the construction companies, they provided the workers, but for every guy with papers they send three or four workers… They were paid 10 euros and they gave 3 to the worker and kept the rest for themselves.[73]

7.3.3 *Comparison*

The work experience of Ecuadorian irregular migrants in Amsterdam and Madrid has presented a number of differences.

In both cities, during the considered years, migrants experienced important changes in their working opportunities.

In Amsterdam, it was possible to recognize two very distinct moments. The first, that lasted until the early 2000s, was characterized by the abundant availability of jobs in numerous sectors (construction, services, industry, cleaning). The second, from the mid-2000s on, was characterized by a progressive reduction of the available sectors. In particular, for irregular migrants it became increasingly difficult to find working opportunities in sectors other than private-house cleaning. As emerged from the interviews, this change was largely due to a restrictive turn on the part of the authorities. The increased inspections on the working sites and the higher fines in case of misconduct made it inconvenient for employers to hire irregular migrants. Moreover, the continuous arrival of large numbers of regular migrants from Eastern Europe offered them a valid alternative. Partly out of necessity, partly as an adaptive solution to the changed scenario, then, irregular migrants progressively moved to the private-house cleaning sector where they found a pretty stable, safe and rewarding labour niche.

Also in Madrid, it has been possible to distinguish two very different phases regarding irregular migrants' working opportunities. The first phase, which lasted until the end of the 2000s, was characterized by a great availability of working opportunities in many sectors. Although this was the case for both men and women, a rather marked sectorial division was registered. Men were mostly employed in the construction sector, women in private-house cleaning and the care sectors. The second phase, which started in 2008, with the beginning of the economic crisis, was characterized by a sharp reduction of the working opportunities that deeply affected

[73] Interview with Walter (16SM).

all migrants. For irregular migrants, in particular, it became extremely difficult to find any occupation. The effects of the economic crisis on irregular employment were made even worse by a stricter control policy and the availability of workers with a regular status. Within this new changed scenario, the opportunities for irregular migrants became very limited. The construction sector, which had been the main attraction pole for men, simply collapsed. For women, the situation was slightly better, because the cleaning and care sectors were less affected by the economic downturn. Many migrants decided to move to another country or to go back to Ecuador. Those who remained tried to survive doing small jobs in the construction sector, transportation, or in services. For men, an option was to switch to the cleaning and care sector.

On the whole, while the double scenario is similar in both Amsterdam and Madrid, the underlying reasons for the dichotomy appear different, and likewise the consequences. In Amsterdam, the causes of changes experienced by Ecuadorian irregular migrants appear to be mainly political, in Madrid mainly economical. In Amsterdam, the increasing number of inspections in many economic sectors caused a sectorial shift on the part of the migrants. Since working in sectors such as construction, services and industry became increasingly difficult and risky, irregular migrants moved to the private-house cleaning sector. In Madrid, the effects of the economic downturn caused a general reduction of the working opportunities. For irregular migrants, it became very difficult to find a job in any sector. The reason in this case, was not, or not principally, that there were more controls, but simply that there was no work at all. Those migrants who had been able to regularize their status and who were also unable to find any employment have confirmed this impression.

Regarding the working opportunities for irregular migrants, two further differences can by underlined. Firstly, in the first phase, Amsterdam displayed a more even distribution of irregular migrants in different sectors (construction, services, industry, cleaning in private houses) and then, in the second phase, a concentration in one (cleaning in private houses). Madrid, instead, displayed a more marked concentration in some sectors (construction, cleaning and care) in the first phase and, in the second phase, a concentration in two (cleaning and care) but with scarce opportunities even there. Secondly, the cases revealed a different situation concerning gender distribution. While in both cases a certain sectorial difference emerged, in the case of Madrid this was much more marked.

An interesting facet regarding the topic under discussion concerns the care sector and, in particular, the care service for the elderly. While this sector has had a crucial role in Madrid, employing a vast number of irregular migrants and especially women, in Amsterdam employment in this sector has been completely absent. As pointed out by many migrants, in the Netherlands, the government offers a number of subsidized services to the elderly so that no working opportunities "under the table" are available in the sector.

Finally, also regarding the working conditions, the two cases have shown a differentiated picture. In Amsterdam, notwithstanding the necessary sectorial shifts, the working conditions for irregular migrants have generally and steadily been valued as positive. The large majority of the interviewed migrants told stories of relative success. They were treated well, were able to save money and to fulfil their economic expectations. In Madrid a distinction must be made. The interviewed migrants clearly distinguished in their stories between the pre-crisis and the crisis period. The first was generally characterized, although with a slightly higher number of exceptions, by a great availability of opportunities, good working conditions and economic success; the second, by very limited working opportunities, unstable and underpaid jobs.

7.4 Internal Controls

In this section, our analysis will be cantered on the experience of internal controls that Ecuadorian irregular migrants had in Amsterdam and Madrid. In particular, two aspects will be discussed: A. the experience of police (or other authorities) controls that migrants had in their daily lives; B. the actual fear that migrants had of being deported migrants.

7.4.1 Internal Controls in Amsterdam

All the interviewed migrants agreed on the fact that in Amsterdam there are no police raids or other authority controls specifically aimed at apprehending irregular migrants. However, they also agree on the fact that every aspect of everyday life, for instance, riding a bicycle, walking on the street, taking a bus, is severely regulated and closely controlled. A small slip during one of these activities can lead to an identity check and, therefore, for an irregular migrant to possible detention and expulsion. Going out at night to clubs, bars and discotheques can also be risky because in these places there can be controls when there are fights, selling of drugs, etc.

> No, no… in this country there are no controls in the streets… I mean, there are controls in the labour sites, as for all workers, but not for the papers. Then if there is something irregular, they can ask you for your papers, but they never come for the papers… Here in the Netherlands, only if there is a complaint or if you made a slip, can they check you… Otherwise no… they let you live in peace… Those who have been caught, it is because they were in a nightclub, they had been drinking too much and they started a fight outside. Others, pitifully, were caught with the bicycles… We had never taken a lesson on how to ride a bicycle… like those for driving a car… This is a bicycle country, everyone moves with a bike. If you make a mistake and you are unlucky, a policeman stops you… Many of us have fallen for a red light, or for other little things… Those little things betray you…[74]

> Here you can do whatever you want, unless you break the law… You can go to a nightclub, you can go to a park, you can get together with friends wherever you want without

[74] Interview with Maria (6NW).

problems… It is not like in other countries that they look at your face, they see a stranger and they ask for your papers… No. This is one of the safest countries in that regard. There is no such thing as them checking you, just in case… They say that in other countries you are walking in the street and they ask you for the papers… There were controls here, a couple of times, but they were looking for guns. They announce these controls, they say: next week there will be controls for guns. They stop you in the metro, check your bag and that's it. They go after guns, they don't go after illegal people.[75]

I know that they know about us… The Dutch know about all the migrants, legal and illegal… Here everything is controlled… You think: I am hidden, they cannot see me… But I think that the Dutch are very smart and that here everything is checked. They know how many migrants are here… They know but they don't do anything… they wait until you fall… and in that moment… For instance if you walk in the street and you cross a red light, the police come and plin!! They get you. Or if you do anything wrong, or make a mistake, any mistake… they can get you and send you to your country. You always need to go around with your eyes wide open.[76]

Irregular migrants in Amsterdam have generally conducted normal lives as regards their free time or their movements around the city. In this sense, no particular feelings of threat or pressure on the part of the authorities was described. As pointed out by many, the important thing in the Netherlands is to respect the law, to avoid committing not only crimes but also "small faults" such as crossing a red light or having an expired ticket on the bus.

I've done everything… football, shopping, nightclubs… I have never felt inhibited… Of course, if you don't do illegal things… If you start a fight also if you are legal, you will have a problem… Imagine if you are illegal. Then, there is also a matter of luck… If you are in the wrong neighbourhood, at the wrong moment, it can happen that there is a raid for drugs and you are checked… But that is very unlikely…

"If you compare how the police work here and in Spain, it is very different. Here in the Netherlands the police are very tolerant. In Spain it is unbelievable… If you look like a *latino*, the police ask you for the papers. If you don't have them… to the jail… I travelled to Madrid once, to try to regularize my status there… The lawyer that was helping me said to me: I will give you a piece of paper sating that you are in the process of getting your papers… it is not legal, but if they stop you it helps. I left his office and took the metro. At the exit, in the underpass, the police stopped me: papers!! I gave them the paper I had just received… They said: ok! Pufff… I could not believe it… Then, I continued… That same day, in another metro station, another check… I said to myself: I have to go away right now. You know… here it is very rare that they stop you… The police do not run after illegal migrants. Unless you get into trouble, the police are calm… In Spain, in contrast, with or without problems, the police ask for papers… I have been here for 13 years, they have asked me for the papers 2 times… In Spain two times in one day…[77]

Here you need to follow the rules, you need to become responsible, you need to become organized. The migrant, the illegal person, cannot go every weekend to the nightclub until the morning… You know, in those places, legal or not legal, the police can ask you for your

[75] Interview with Juan (10NM).

[76] Interview with Lurdes (25NW).

[77] Interview with Mauricio (18NM).

documentation… You have to be careful… Especially if you have a family… Look, if they get you, you will be back in Ecuador! Period! What will your wife do? What will your children do? So, what do you prefer? To make a sacrifice for your family or for your friends? For your family![78]

Regarding the perception of the risk of being deported, the experience of the Ecuadorian migrants in Amsterdam has revealed an increasingly severe scenario. While expulsion has always been a concrete risk for irregular migrants in the Netherlands, since the early 2000s it has become almost a certainty for an irregular migrant who is caught.

There was a change around 1999 or 2000… Before that, even if they caught you, it was rare that they deported you… But, after that, they started deporting everyone… We were scared, you heard about this and that… Many people who were deported before 2003 were able to come back a week later… In Ecuador, you simply ask for a new passport… They had a different number each time… So, there was no problem… But after 2003, they started asking for the visa… If they sent you back, you could not return.[79]

Here there are no controls in the streets. Absolutely none! And I agree, there should not be… If they do that… we will be going back to the second world war… to Hitler… He made that kind of controls on the people… I think that this country suffered a lot due to that situation and it is because of that, that today they care a lot about human rights… What I don't like here is that they deport you for things that are not… not really important. They take as an excuse for instance, if you use the tram without a ticket, or if you go by bike with a broken light… They get you and they deport you. They need to justify the deportation… they need to respect the law… If you are robbing the state, they have a good reason: a migrant cannot rob the state! So they find an excuse and they deport you.[80]

Controls have become more intense each day. Every year they change the laws because they don't want people in this country anymore. It is getting harder all the time… I think they are angry because the migrants abused everything they gave them, especially the Turks and the Moroccans… It happened, for instance, that they were not working here but since they had social benefits, they could maintain the children they had back in their countries… I think that is what offended them… so now they are doing everything to eliminate migration, to make the life of the migrant impossible… especially that of the illegal migrant… You cannot study, you cannot get your qualifications, you cannot go to the hospital, etc. That type of thing…[81]

If they get you and are able to know who you are, they kick you to Ecuador 100%… There is no way they will let you go now… Two nephews of mine have been deported… A friend of mine too… You know… It is very tough… So, you should not carry your documentation… If they catch you, you have to give the name of a legal person…[82]

[78] Interview with Pablo (29NM).

[79] Interview with Lorena (26NW).

[80] Interview with Gabriela (27NW).

[81] Interview with Javier (23NW).

[82] Interview with Mauricio (8NM).

7.4.2 *Internal Controls in Madrid*

The experience of internal controls by the Ecuadorian irregular migrants in Madrid can be clearly differentiated into two phases.

While random identity checks and controls in the streets have always been a possibility, until the second half of the 2000s, these were very limited, unsystematic and largely inconsequential. A migrant could be stopped, asked for papers, even taken to the police station, yet this was very rare and usually did not have major consequences.

> I was always outside... nothing happened... When I was illegal, the police stopped me on two occasions... The first time I was waiting for the bus... A policeman came and asked me for papers... I had only a photocopy of my passport... I was scared, but nothing happened... He looked at the picture, looked at me... and said: it's ok! Don't get into trouble... And he left... The other time was the same...[83]

> I did not have a residence permit for more than 4 years... but I was not afraid I have to say... You know, the first months that I was here, I was worried but then I realized that there was no problem... There were no controls, there was nothing... I mean, probably there were controls here and there, just to take a picture and say that they were controlling, but c'mon... They all knew where we were. You just had to go to *Tetuán* or to *Casa de Campo* [a neighbourhood and a park where Ecuadorian migrants used to meet] and you could meet the whole of Quito [the Ecuadorian capital] playing football... If they had wanted, they could have sent 200 buses, got the people on a plane and sent them home...[84]

Since the beginning of the economic crisis in 2008 and, in particular, after the change of government in 2011, a marked change has taken place. Controls in the streets, metro stations and in gathering places, such as, parks, bars and discotheques became much more common, although intermittent. The migrant spoke about "spells of controls": particular months or weeks in which the controls increased.

> This government that has entered now is always complaining about the migrants... the migrants, the migrants... The other, instead, Zapatero, I think he was in favour of the migrants, but this, this one hates us... You can see by the controls in the streets... Now the people without papers are afraid... Now you don't go out even to look for a job...[85]

> One day I was in the metro... I was listening to the headphones... Two guys approached me... I thought they were going to sell me something... Then they took out their badge... They were policeman in plain clothes... They said: papers! Luckily it was all right... I had my papers...[86]

> A couple of months ago, there were many controls... it goes in spells. In Metro Plaza de Castilla [the name of a metro station], there were those *paisanos* [policemen in plain clothes] as they call them. There were many of them, checking for papers. And also in *Metro Usera*... I always go there, because one of my cousins lives there... And I saw that

[83] Interview with Hernán (24SM).

[84] Interview with Daniel (7SM).

[85] Interview with Patricia (1SW).

[86] Interview with Lucy (15SW).

> they were also stopping women… they were taking them away… I could not believe it… Because, before, they used to stop only men, but now also women…[87]

> Now there are a lot more controls. On two occasions they took me to the police station… In the two cases we were playing football with some friends… And you know, after the football, we always buy some *litronas* [beers]. After a while, the police came and started checking our papers… I had left my wallet at home… So they took me to the police station… Luckily that was when I still had the residence permit… But they kept me there from Friday to Monday… only because I did not have the documents with me… Right now they always come to check us… to the parks, to the little squares… when they see a group of us, they come and check our documentation. Also in the metro, there they are often in plain clothes.[88]

> Luckily I got the papers in 2008 because, after that, they started with the raids… In every corner, in every metro station they could stop you. We were paranoid… I think it was in relation to the crisis… It was in that moment that the pressure against the migrants started… And especially against the migrant without documents. What they were trying to do was to scare the people so that they wouldn't come anymore.

> When the raids started, there were *latino* radio stations that alerted the places where the police were… They alerted the illegal migrant… Be careful in that station, they are checking for documents there… It was the people who called to say where the controls were…

Also regarding the perception of the risk of being deported, the Ecuadorian migrants in Madrid have distinguished two phases. Until the beginning of the economic crisis, it was very unlikely for an irregular migrant to be deported. Most of the interviewed migrants agreed that it was very infrequent that you heard that someone had been sent back to Ecuador. After 2008, this situation slightly changed. As the controls increased, also the possibility to be deported increased. Yet, if in this second phase, those who were still in an irregular situation started to be more worried about a possible deportation, all the interviewed migrants agreed on the fact that, in order to be deported, it was generally not enough to simply not possess the residence permit.

> Here in Spain it is not that they get you without papers and they deport you. No! You need to have a criminal record… You need to have been involved in something like fights, thefts, vandalism… Otherwise it is very difficult… They can even take you to jail or the CIEs [administrative detention centres] but they let you go after a while…

> For them to take you to the CIE, you need to have done something, to have been involved in some trouble. A change took place with the beginning of the economic crisis. From that moment on, they started to check and deport the people. I had been stopped before, but, I swear, it was as if they checked you only to check you… The second time, I think it was two years ago [2011], a policeman stopped me. You could see that it was not like before. Now, they stopped the people and sent those without papers away. Now they were checking in order to deport.[89]

[87] Interview with Pilar (6SW).

[88] Interview with Pablo (4SM).

[89] Interview with Juan (10SM).

> I lost the papers in 2008, 5 years ago… It was because of a fight I had had with a policeman. You know in this country there is a lot of racism… They never check the Spanish kids… they go to the same park or to other parks… they drink, do drugs, pee on the street… everything. But if the police arrive, they come directly to us… They see a group of *latinos* and they come. They treat us very badly, they put us against the wall to check our pockets. So I made the mistake of reacting… The policeman pushed me, I pushed him, he gave me a punch, I did the same… After two seconds I had all the policemen over me… I had a trial and of course I lost, there were 15 policemen accusing me… In that moment I lost the papers… Since that moment it has been hell… I have been to the CIEs 4 or 5 times… Every time they stop me, they ask me for the papers. I end up in the CIE. They have not deported me because my father always sends a lawyer… If you don't' have a lawyer that acts within 5 hours and you have a criminal record like me, they put you on a plane… If you don't have a criminal record and you have someone who helps you, they don't send you".[90]

> Here the police are very nice… If you do what you have to do, if you don't get into trouble they don't do anything to you. Here everybody is complaining about the police, but if you pay attention, those who complain are those who have done something. I don't fear the police… I always think that I am not a criminal, that I have nothing to be ashamed of. Many times they have asked for the papers. Many times. I tell them that I am in a process, that I am waiting and that everything is going to be all right. They always say to me: ok! Good luck! If you are serious and explain your situation, nothing happens. They deport the criminals, not those who are not doing anything. If they get you with drugs, drunk in a nightclub, in a fight, they put you on a plane…[91]

7.4.3 Comparison

The results that emerged from the fieldwork have revealed two different situations regarding internal controls in Amsterdam and Madrid.

In Amsterdam, there have not been *ad-hoc* controls on irregular migrants in the streets or in public places. Migrants, therefore, did not feel under direct threat and did not usually feel scared about moving around and carrying out their normal activities. However, the rigid checks regarding respect for the rules that regulate most social activities, such as, walking in the street, riding a bike, using public transportation, have been an indirect form of control. Irregular migrants know that a simple mistake, an administrative fault of any kind, can lead to an identity check, to administrative detention. For these reasons, these people are usually very alert to the situation around them at all times and very self-controlled in their public activities. Regarding the possibility of being deported, the impression gathered is that in Amsterdam this has become in the last decade a very realistic one among irregular migrants. In other words, most migrants seem to know that if they get caught, the most probable consequence is that they are going to be deported.

In Madrid, there has always been the possibility of *ad-hoc* controls on irregular migrants. Until the second half of the 2000s, though, these were very limited,

[90] Interview with (19SM).

[91] Interview with 20SM.

unsystematic and largely without any consequences. Migrants, therefore, described a very relaxed situation and a negligible possibility to be deported. After the start of the economic crisis, this scenario changed. The controls on irregular migrants became more frequent and systematic. Every street, metro station or public place could be the place for a potential raid. These controls, however, were rather intermittent. They increased in particular months or weeks and diminished afterwards. While the fear of being deported certainly increased in the second phase, for the interviewed migrants, this possibility remains rather unlikely. As pointed out by many, in order to be actually deported, it is usually not enough to simply not possess a residence permit; the irregular migrant has to have a criminal record.

7.5 Housing and Healthcare

In this section, the experience of Ecuadorian irregular migrants in Amsterdam and Madrid regarding three important aspects of their daily lives, namely housing, healthcare and education for the children, will be analysed.

7.5.1 Amsterdam

Housing

Finding a house to live in and stay for a relatively stable period has been one of the most difficult tasks for irregular migrants in Amsterdam. All the interviewed migrants, with no exception, indicated housing as the biggest problem they had to deal with in Amsterdam. The difficulty was related to the general scarcity of houses and to the existence of a very controlled system of public houses. The main available option for irregular migrants was to sublet rooms or entire houses from people who get the houses from the public service. This option, however, was usually very unstable because this type of houses is greatly controlled by the authorities. If a control came, the migrants had to depart in that moment, many times leaving their belongings behind, or losing the money they had paid to get the house.

> The house is the biggest problem here in Amsterdam. Imagine: there is no house for the Dutch, so what do you expect for irregular migrants? I have changed more than 10 houses in the last number of years… It is really bad.[92]

> Here it is very difficult. If you don't have someone who knows you, it is very difficult to get a good house. You cannot go and say: I want a room or a house. They ask you for your residence permit. The other option is to sublet a room in a government house, but in that case you never know if a control may come. Those houses are very much controlled. They come

[92] Interview with Marta (7NW).

to check if the owner is really living there or if he is subletting. That is very common here in Amsterdam, someone gets a house from the government, but they don't live there, they go to live with their girlfriend or boyfriend, and rent the house under the table… You are always scared in that situation… Unless the owner lives there as well, you'd better be careful…[93]

Every Ecuadorian here has struggled for the house… everyone. It is horrible, horrible! I have lived in 15 different places… I think the only district where I have not lived in is Amsterdam-Noord; otherwise I have lived everywhere. The problem is that the country is small and there are a lot of people. Even for the Dutch, it is difficult to get a house. They have to wait 7, 8, 15 years in order to have a house [she means a public house]. The houses are very small and expensive. So, if you rent a house or a room, you have to pay the deposit, one month in advance and the first month. After one month they can tell you: Out!. And you cannot do anything. The money is lost![94]

If you have money it is easier to find a house. Here, there are a lot of people who work as intermediaries. The risk of fraud is high. Once we lost 2,000 euro in 10 minutes. They tell you: today I have three clients that want this house. They know you are illegal… If you get the house, when they want, they can come and say: get out or I'll call the police… If you are lucky you can stay in the house for 2, 3 years…[95]

The house is the most difficult thing here… I have been here for 13 years and I think I have changed 24 houses. That is because I have always found a house with people that have a government house and those are the people that are checked most by the police. So, every minute a control can arrive. We have been in a house for 3 months and out, three months and out! Sometimes you don't know what to do… With my husband we have slept in the park… Once, with my child, we had to sleep in a taxi. A guy from Suriname was very kind and he let us sleep in his taxi… The guy where we used to live let us the room for 600 euro… He rented the other room as well. He occupied the living room. I think for the whole house he had to pay 400 euro… Do you understand? He made a lot of money. For sure someone told the police he was subletting. We always had to leave the houses for this kind of situation… people who were living on public benefits or who had some kind of trouble. If a letter arrived that the control was coming, we had to leave.[96]

Healthcare

The access to healthcare for Ecuadorian irregular migrants in Amsterdam has been another problematic issue. If, until 1998, they had been able to freely access the public service, from that year on, the residence permit became a necessary requirement to be treated for free. As a matter of fact, no healthcare insurance could be stipulated without a valid residence permit and, without insurance, the people had to pay all medical assistance. For irregular migrants, consequently, the only option

[93] Interview with Marco (11NM).

[94] Interview with Gabriela (27NW).

[95] Interview with Pablo (28NM).

[96] Interview with Laura (24NW).

in order to access the public medical assistance was to pay. Many migrants had medical problems, they all went to hospital and paid.

> For more or less 10 years now, things have been very difficult. We were not 50 or 60 who went to the hospital… We were 500… Each time that you go to the hospital it costs at least 300–400 euros… So that was a debt that the government has. I think it was for that reason that they changed the law. Now you cannot buy a medical insurance, you need the residence permit. There is a fund that the hospitals can use in cases of emergency for us… [by "us" she means irregular migrants]. So now, what we have to do is to hope not to get ill, otherwise you have to go and pay the bill… Sofia, that friend of mine, broke her leg… I think she has 15,000 euro of debt with the hospital…[97]

> We solved the healthcare issue by eating well, so that we don't get ill… If we get ill we use "home medicine"… The hospital is too expensive and since we have no medical insurance it is also a bit difficult… Once I broke my elbow… I went to the hospital and I needed x-rays and a head scanner, because I had fallen down the stairs the stairs. They thought I could have something in my head… It happened while I was working in a house. The woman said to me: are you ok? I am sorry!! She didn't say anything else… She didn't ask how much I had to pay in the hospital… When you don't have papers, you have no rights… So I went to the hospital… In the end, a social worker came and said to me: you have to pay, what do you want to do? She told me that I could pay in monthly instalments. For the whole thing it was 2800 euro. I had to pay, because there are others who don't pay, they use the fund that the hospital has for irregular migrants… The problem is that if you do not pay, the next time they don't see you. Now they have a file with my information. If I go again they know everything, even my form of payment… Now, for instance, I will have to go to give birth [Luisa is pregnant at the moment of the interview], I don't know how much I will have to pay…[98]

Some migrants have internalized this situation and act accordingly. Mauricio, for instance, has a very pragmatic understanding of the irregular migrants' relations with the state and the public services. If you have a residence permit, you can use the insurance system; if you do not have insurance, you have to save the money and be prepared to pay.

> When you are illegal, you have the advantage that you earn more… If you work legally, you have to pay a lot of taxes; they take out 400 euro, 500 euro per month. But, if you think logically, when you get ill you don't have insurance, when you are old you will not have a pension. So the extra money you get now, you have to save it. You have to build your own insurance under the mattress…[99]

A number of alternative strategies have been elaborated to avoid the high costs of the public healthcare. Some migrants have been able to use the insurance number and the documents of other people. Others, in the case of small problems, have gone to the Red Cross, which provided a basic service for irregular migrants under the payment of 5 euro. Another option was to go to private "unofficial" doctors who worked under the table. Finally, an interesting case was that of Gabriela, who, to avoid the costs of giving birth in Amsterdam, decided to travel clandestinely to Spain.

---

[97] Interview with Maria (6NW).

[98] Interview with Luisa (29NW).

[99] Interview with Mauricio (18NM).

Luckily I never had problem, so I didn't need to go to the hospital. I discovered how expensive it was when I had to give birth. One echography costs 200 euro… Here they don't give you anything, absolutely anything if you are illegal. Now I am legal and I pay insurance, so I can go whenever I want. But when I was illegal I could not. I mean, I could but I had to pay. You can go and give birth but it can cost 7,000 – 9,000 euro. When you are a first-time-mother you often have complications… and the price rises. I had complications!! I thank the Lord I went to Spain to give birth, because here it would have been 10,000 euro. I would now have a debt… My mother used to live in Spain, she told me to come here to give birth. I decided to go. I have to say that they often say that people are racist in Spain, but I think that, regarding healthcare, they are very humanitarian. I went, I gave birth, they treated me, they cured me. I stayed in hospital for 5 days. Can you imagine how much that costs here? Moreover, here they treat you very badly… They kick you out of the hospital the same day… If you have a complication, you stay 5 hours, not like in Spain or Ecuador where you stay 5 days.[100]

7.5.2 Madrid

Housing

Finding a house in Madrid was a problem for an irregular migrant only in the very first years of the Ecuadorian migration. As pointed out by many of those who were part of the first wave of migrants, at the end of the 1990s, there were not many houses and it was not easy to find accommodation. In this early stage, the most common solution was to rent a room or even a bed in a room from people who were making money from this kind of business. The conditions were usually very bad and the prices relatively high.

When I had just arrived [1997] I didn't know anybody… I had a cousin who told me that there was a Peruvian woman renting rooms… I remember that I went to talk to her and she told me the price… I don't remember because it was in *pesetas*… I think it was something like 190 euro… The room was big… And so, I asked if the people who owned all the luggage that was in the room were coming to pick it up. She looked at me as if I was crazy… I thought the room was all for me… [laughs]. There were 8 of us in that room, can you believe that?[101]

At the beginning it was difficult. The Spaniards didn't want to rent you a house without the *nomina* [a working contract] and you could not have a contract without the papers… So the only option was to rent rooms… You know, in Ecuador there is a lot of space… we were not accustomed to renting rooms, to living with other people that you don't know… It was hard.[102]

A friend helped me when I first arrived… They had a small house, because he had already got papers… I had to sleep on the couch… the problem was that they rented the other couch to another women from Ecuador and they had a big dog… I was crazy! After one day I said: I cannot stay here… They helped me to find another place with a friend of theirs… it was

[100] Interview with Gabriela (27NW).

[101] Interview with Xavier (5SM).

[102] Interview with Hernán (24SW).

> on another couch… That was at the beginning… Then you start making contacts… knowing people and it becomes easier. The next place was an entire room for me in a house with other migrants…[103]

As the first migrants started to regularize their status and were able to rent entire houses, the renting market rapidly expanded. Already in the first years of the 2000s, the issue of the house had become a lot easier for irregular migrants. A great number of Ecuadorians, moreover, helped the newly arrived to rent from family or friends.

> For me the house has not been a problem… When I arrived I stayed at my sisters', I think for 2 years… Then my boss told me he had a flat… You know… they were building entire neighbourhoods… For whom do you think they were building all those houses?? For the migrants who were arriving in hordes!! So I went to live in that house… My boss did not ask for papers, of course, he knew I did not have them, but then when I got them, he gave me a contract. And do you know what I did with the other room in the house? I started renting the other room to a friend of mine…[104]

> In the first years the housing situation was not easy… I think it was because the Spaniards did not trust the Ecuadorians… Then I think they started to know us better, to see that we were good workers. As soon as I got the papers, I was able to rent this house. I have been living here for more than 11 years… The owners are very happy with us because we never missed a payment and we don't create problems. My house has been like the gate to Spain for the Ecuadorians… I have hosted all my family, friends, friends of friends…[105]

Healthcare

Healthcare has not been a problematic issue for Ecuadorian irregular migrants in Madrid. As established by the law, all the migrants registered on the municipal record, with no regard to their administrative status, were allowed to freely access the public healthcare system. All the interviewed migrants have confirmed the correct implementation of this provision; none of them has had any problem in accessing the service.

7.5.3 *Comparison*

The access to housing and healthcare for Ecuadorian irregular migrants in Amsterdam and Madrid has presented a number of important differences.

Regarding the first issue, the access to housing in Amsterdam has been generally much more difficult than it is in Madrid. This has been due to two main factors: A. the lower supply of housing opportunities; B. the existence of a strictly regulated and controlled system of public housing. The combination of these factors

[103] Interview with Lorena (25SW).

[104] Interview with (26SM).

[105] Interview with Leticia (11SW).

determined a very precarious situation for irregular migrants' housing. The available options were generally unstable, expensive and at risk of frauds. In contrast, in Madrid, after a first moment in which the housing opportunities had been relatively scarce, the situation rapidly improved. As many Ecuadorians and migrants from other countries started to get their status regularized, they were able to rent entire houses or flats and sublet rooms to the newly arrived. This determined a quick expansion of the housing opportunities and, therefore, the availability of relatively cheap, stable and safe housing for the irregular migrants.

As far as the second issue is concerned, a similar situation has been found. In Amsterdam, access to healthcare has been much more problematic for irregular migrants than in Madrid. In this case, the determining factor was the different regulations regarding access to the public healthcare system. Whereas in Amsterdam, irregular migrants have been excluded from non-emergency care since 1998, in Madrid they could freely access the public system until 2012. Hence, while for irregular migrants in Madrid, the issue of healthcare was basically not a problem, in Amsterdam they had to find ways to overcome the existing limitations. The most common option, in case of serious medical problems, was to go to the public hospitals and pay the costs at market prices. For minor problems, there was the option of medical support offered by humanitarian associations or by private unofficial doctors.

7.6 Irregular Migration Realities in Amsterdam and Madrid

The first and most important conclusion stemming from the fieldwork realized in Amsterdam and Madrid, is that the experience of Ecuadorian irregular migrants in the two cities has been radically different. The diverse combination of possibilities, limitations, opportunities, resources, etc., present in the different spheres of social life in the two contexts determined a set of very different conditions for irregular migrants in order to develop their lives and fulfil their objectives. This result seems to confirm the hypothesis which emerged in the theoretical part of this study: the existence of different "irregular migration realities". While in legal terms, the lack of a residence permit generates, in principle, a similar condition, the "social translation" of this condition sharply differs, depending on where such translation takes place.

While this conclusion may seem rather obvious or predictable, the truth is that it is not. As extensively discussed in the first part of this work, one of the main limitations in the current understanding and study of irregular migration has been the tendency to treat it as an undifferentiated phenomenon. It was also saw, how this problem had both a theoretical and an empirical origin. On the one hand, irregular migration had been studied using unsophisticated theoretical tools; on the other hand, there has been a lack of comparative analysis of the phenomenon. If in the theoretical chapters it was tried to challenge this problem in analytical, logical terms, producing the hypothesis of differentiated, systemic contingent "irregular migration realities". Here it was tried to go to the field and discover if such hypothesis was realistic.

In the next chapter, an attempt to assess possible systemic relations between the structural characteristics of the two contexts, analysed in Chap. 6, and the irregular migration realties, which emerged from the fieldwork will be presented. Before that, then, in this final section, an effort to produce a general characterization of the irregular migration realties in Amsterdam and Madrid will be made.

Not only has the experience of Ecuadorian irregular migrants been different in Amsterdam and Madrid, but it has also gone through different phases within each of the two contexts during the considered period of time (1997–2013). This reveals how the irregular migration phenomenon differentiates across space but also across time.

In Table 7.1 it is possible to observe a summary of the main findings of the fieldwork.

The experience of Ecuadorian irregular migrants has offered many insights into the irregular migration realities in Amsterdam and Madrid. Let's see them in detail.

Table 7.1 Irregular lives Amsterdam and Madrid

		Amsterdam	Madrid
Legal trajectories and regularization channels	**Main trajectories**	Never regular (17/39) Regularized through marriage or cohabitation agreement (11/30) Regularized under exceptional circumstances (2/30)	Regularized using ad-hoc channels (21/30) Befallen irregularity (6/30) Never regular (3/30)
	Irreg. years	12 years (average)	5 years (average)
	Channels	Indirect: (1) marriage and cohabitation agreements	Direct: (1) extraordinary regularization programs, (2) through labour quotas; (3) through rootedness
Regularization strategies		Bogus marriage or cohabitation agreement with a Dutch citizen – very difficult Bogus marriage or cohabitation agreement with a EU citizen – difficult	False contracts to apply for regularization False police record to apply for regularization

(continued)

Table 7.1 (continued)

		Amsterdam	Madrid
Work	**Sectors**	Before mid2000s, men: hotels, restaurant, construction, port, and industry Before mid2000s, women: hotels, restaurants, office cleaning, private-house cleaning After mid2000s, men: private-house cleaning and construction After mid2000s: women: private house cleaning	Before 2008, men: construction, restaurants, industry, storage, transportation, and couriering Before 2008, women: private-house cleaning, office cleaning, care work with children and the elderly. After 2008, men: "little jobs" in construction, transportation, care work with the elderly After 2008, women: care work with children and the elderly, private-house cleaning
	Conditions	Work availability: Before mid2000s: high After mid2000s: medium Working conditions: generally medium to good Wages: generally high	Work availability: Before 2008: very high After 2008: very limited Working conditions: generally medium to good, some cases of exploitation Wages: Before 2008: high in certain sectors, medium in others After 2008: medium in certain sectors, low in others
	Controls	Before mid2000s: medium After mid2000s: high	Before mid2000s: very low After mid2000s: medium
Internal controls	**Street controls**	No street *ad-hoc* controls for irregular migrants	Before 2008: limited, unsystematic, inconsequential After 2008: in spells, systematic, with consequences
	Fear of deportation	Before 2000s: medium Until mid2000s: high After mid2000s: very high Who can be deported? Everyone	Before mid2000s: very low After mid 2000s: medium Who can be deported? Those who have a criminal record or precedents
Housing		Very difficult, unstable and expensive	Before 2000/2001: difficult and expensive After 2000/2002: increasingly easy and inexpensive
Healthcare		Until 1998: free access to healthcare After 1998: access under payment with the exception of emergencies	Until 2012: free access to healthcare After 2012: access under payment (no cases in my sample)

7.6.1 Amsterdam

In Amsterdam, the experience of Ecuadorian irregular migrants became increasingly difficult along the considered period of time. The policy changes adopted since the end of the 1990s, appear, in this sense, to have been effective in restraining the living opportunities of irregular migrants. This has influenced both the size and the conditions of the irregular migrant community. While numbers have never been big, it seems that the increasing restrictiveness has determined an effective disincentive on new arrivals.

Irregularity has appeared as a long term, hard to change condition for migrants in the Netherlands. The lack of *ad-hoc* regularization channels and the strict control over possible alternative regularization strategies, such as, marriage and cohabitation agreements, have made it very difficult for irregular migrants to obtain a residence permit. It was, then, normal to find migrants with more than 10 years of irregular residence in the Netherlands. Those few who were able to regularize, achieved this result after different attempts and after investing important quantities of money.

Regarding the experience of internal controls, the scenario has been contrasting. On the one hand, the absence of *ad-hoc* police controls of irregular migrants in the public spaces has generated among irregular migrants a feeling of relative tranquillity and sense of freedom. On the other hand, the strict control over the respect for the rules regulating many social activities, such as, work, house rental, car driving, bicycle riding, street circulation, public transportation use, as well as the strict application of the deportation policy, generated among migrants an ever present sense of vulnerability and a highly developed sense of alert and self-control.

The working opportunities and conditions for irregular migrants in Amsterdam, have been evidently affected by the restrictive turn adopted by the government since the end of the 1990s. The most evident result of this change has been the reduction of sectors where irregular migrants were able to find employment. In the last few years, private-house cleaning has been the niche where most migrants have found stable and remunerative opportunities. That being said, the economic success has been one of the most valued aspects of irregular migrants in the Netherlands. Notwithstanding the difficulties related to the language, the high levels of controls, the risk of deportation, and, in the last few years, the reduction of working opportunities, migrants were generally able to find jobs, to earn and save money and to send money back to Ecuador.

Regarding the access to vital resources, such as, housing and healthcare, the experience of irregular migrants in Amsterdam has been very problematic. As for housing, the particular characteristics of the Dutch housing market and, especially, the limited supply of opportunities and the strictly controlled system of public housing, has determined very precarious, unstable and expensive conditions for irregular migrants. Moreover, the combination of a low supply of and a high demand for houses has fostered cases of frauds and abuses. In the case of healthcare, the impossibility to freely access public healthcare since 1998, severely complicated the situation for irregular migrants. The requirement to pay for the services in public hospitals, implied for migrants the search for alternatives, such as, the help of private doctors or humanitarian associations, the accumulation of debts and, in certain cases, the neglect or mistreatment of dangerous illnesses.

The combination of all these conditions has created an increasingly difficult environment for irregular migrants. As an adaptive solution, migrants have been obliged to develop sophisticated strategies to overcome limitations and barriers or to recur to the services and options offered by underground or even criminal organizations. While the possibilities to regularize have been very limited and many spheres of daily life quite problematic, the economic opportunities have been a sort of counterweight. Although the indirect police pressure is high and the risk of being deported is tangible, Ecuadorian irregular migrants have been quite effective in adapting to the environment and in developing reasonably serene and successful trajectories. Once they learn how to deal with the main problems and to behave in a discreet way, it is possible for them to conduct a parallel existence to that of the "regular" citizens. Paradoxically, in purely economic terms, their options may be even better. At least in the short, medium term, then, most migrants considered their experience as successful. The issue of the papers becomes truly critical for irregular migrants only when their children approach legally adult age and face the possibility of having to abandon their studies and start working with them.

7.6.2 *Madrid*

In Madrid, the experience of Ecuadorian irregular migrants went through two very different phases in the considered period of time: the first, between the end of the 1990s and 2008, and the second, afterwards. The first phase was characterized by the massive arrival of irregular migrants, the abundance of working opportunities and the possibility for the irregular migrants to easily regularize their status. A sharp reduction or even the inversion of the fluxes, the collapse of the job market and the reduction of the regularization opportunities characterized the second phase. Although a number of political reforms implemented by the authorities through the 2000s may have certainly effected the situation, the decisive factor in determining the change of scenario in 2008 was the start of a serious economic crisis in Spain.

Irregularity has appeared as a transitory condition for migrants in Spain. The existence of a number of ad-hoc regularization channels has made it easy for irregular migrants to obtain a residence permit. The crucial role of holding a working contract or a job offer in all the regularization schemes, however, made it a lot more difficult for irregular migrants to get a residence permit after 2008.

Notwithstanding this, at least among Ecuadorians, it has been very difficult to find migrants with more than 5 years of irregular status.

Regarding the experience of internal controls, the scenario has been very different in the two described phases. While police controls on irregular migrants in the public spaces have always been carried out, until 2008 these were very limited, unsystematic and usually inconsequential. After 2008, this type of control was implemented in spells, but in a much more extensive, systematic and determined way. In connection with this development, the perception of the possibility of being deported, which until 2008 had been negligible, definitely increased. Yet, as underlined by most migrants, even during the apex of police raids and deportations in 2011 and 2012, the common perception was that in order to be expelled, it was necessary to have a criminal record or a recurrent story of detentions. As for other types of controls, such as those on the working sites, house rental and other social activities, although migrants have perceived a restrictive trend, especially in the second part of the 2000s, these controls have not been a reason for major concern. On the whole, then, it is possible to say that Ecuadorian irregular migrants in Madrid had a very serene and carefree experience of controls until the end of the 2000s. After that, the concern increased and, in particular, during certain periods, it created a concrete sense of vulnerability.

The working opportunities and conditions for irregular migrants in Madrid, have been evidently affected by the economic crisis which started in 2008. If, until then, both men and women had been able to find plenty of opportunities in a number of sectors, after that, especially for men, it became truly difficult to find an occupation. This dramatic change severely affected the economic situation of all migrants and particularly of those without a residence permit. The sharp increase in unemployment among the general population, made it very difficult for irregular migrants to get a job offer. The only sectors where they were still able to find relatively stable and remunerative jobs were private-house cleaning and the care sector. The assessment of the economic experience of Ecuadorian irregular migrants in Madrid can be considered as twofold. Until 2008, these irregular migrants were successful because they were able to work, earn and save money, and also send money back to Ecuador. After 2008, these people were unsuccessful and in most cases had to limit themselves to basically surviving.

Regarding access to vital resources, such as housing and healthcare, the experience of irregular migrants in Madrid has been relatively easy. As for housing, the rapid expansion of the market and the low level of controls in the sector meant good, inexpensive and stable opportunities for these migrants. As regards healthcare, the possibility for irregular migrants to freely access the public healthcare system generated a very convenient and unproblematic situation.

The picture that emerges from this overview once again is twofold. Ecuadorian irregular migrants in Madrid experienced two very different phases. Although each phase was characterized by a number of specificities, the most decisive, discerning element appears to have been the difference in the working opportunities. After 2008, the deterioration of the economy severely affected all the population. However, the effects of the lack of work were particularly relevant for irregular migrants since this not only affected their economy, but also their possibility to regularize or maintain a temporary residence permit.

Part III
Conclusion

Chapter 8
Steps Towards a Systemic Theory of Irregular Migration

In this last, conclusive chapter, the main question at the origin of all the research work in this book – how can irregular migration be explained? – will be the focus of the discussion. The chapter is divided into two parts. In the first part, the aforementioned question will be addressed in relation to cases of the empirical study and it will then be reframed in the following terms: how can Ecuadorian irregular migration in Amsterdam and Madrid be explained? To answer this question, the efficacy of the "classic" theories, discussed in the Chap. 3, will be firstly tested. As it will be pointed out, many of the limitations which emerged in the theoretical discussion will become evident also when those theories are applied to concrete cases. Then, a systemic explanation, based on the theoretical approach developed in Chap. 4, will be proposed as a possible alternative.

In the second part of the chapter, bringing together the results emerged from the different research strategies presented in the book, the systemic theory of irregular migration outlined in Chap. 3 will be further developed. In particular, a systemic analytical framework for irregular migration will be proposed. Such framework should be considered as an initial, and necessarily perfectible, attempt towards the construction of a general tool of analysis of irregular migration as a structural, differentiated phenomenon of contemporary world society.

G. Echeverría, *Towards a Systemic Theory of Irregular Migration*, IMISCOE Research Series, https://doi.org/10.1007/978-3-030-40903-6_8

8.1 Explaining Irregular Migration in Madrid and Amsterdam

8.1.1 Ecuadorian Irregular Migration in Amsterdam and Madrid: The Weaknesses of "Classic" Theoretical Explanations

The results emerging from the empirical study have clearly shown that Amsterdam and Madrid display a very different picture regarding the characteristics of structural context that enabled irregular migration, Chap. 6, and also the characteristics of the irregular migration realities that in relation to such context have developed, Chap. 7. This empirical result, that, from a certain point of view, may appear rather obvious, is, however important not only because it confirms on a solid basis what previously was "only" obvious, and this is an underestimated function of research, but most of all because of the theoretical implications involved. As was extensively discussed in the first part of the book, and in particular in Chaps. 3 and 4, one on the main limitations of most theories developed to explain irregular migration was that it was treated as a single, undifferentiated phenomenon. The explanation for irregular migration put forward for a particular case, and sometimes even for a particular case during a limited time span, was proposed as a general, universal explanation of the phenomenon. The confirmation, then, that irregular migration is a differentiated phenomenon, emerging in different contexts and displaying different characteristics, challenges such theoretical assumption and demonstrates the need for a more a sophisticated and versatile theory.

The experience of Ecuadorian irregular migrants in Amsterdam and Madrid not only appeared to be very different in the two cities, and therefore in relation to the geographic location, but it also depended on the moment in which the phenomenon was considered within each context, and therefore in relation to its chronological position. In brief, in Amsterdam, during the time considered: the number of irregular migrants continuously fell; the possibility to regularize was severely limited, thereby, marking irregularity as a long-term status; controls became increasingly sophisticated and pervasive; work opportunities progressively diminished so that, in the last years covered in this study, only few openings, for instance in the private houses-cleaning sector, were left for irregular migrants; access to housing and healthcare became increasingly difficult. In Madrid, two very different phases were discernible: the first, until 2008, witnessed high numbers of irregular migrants; the presence of accessible and effective channels of regularization which marked irregularity as a short-term, transitory status; controls were very limited; there were copious and diversified work opportunities; it was extremely easy to have access to both housing and healthcare. The second phase, from 2008 onwards saw a drastic reduction of the irregular migrant population; a reduced but still present availability of regularization channels; increased yet unsystematic controls; very limited work opportunities; easy access to housing and healthcare. Evidently, this very differentiated picture calls into question all those theories of irregular migration that are unable to account for such geographical and chronological variance. Indeed, all those theories, fail to be consistent with a key precondition of every theoretical effort: its ambition of universality.

Yet, even closing an eye on this crucial limitation, other problems seem to be discernible. Notwithstanding the great number of hypotheses proposed to explain irregular migration, in Chap. 3 (see in particular Table 3.2), two main, broad, alternative underlaying arguments were identified. On the one hand, there is the idea that irregular migration is the result of a diminished ability (if not the complete failure) of states in their ability to control populations. On the other hand, there is the idea that irregular migration is the result of a state choice, adopted to attain, through the manipulation of populations, a number of possible goals. What explicative ability have these two broad arguments if applied to the experience of Ecuadorian irregular migrants in Amsterdam and Madrid?

A first point to stress, which concerns the universality problem just highlighted, is that even if one of the two alternative arguments was effective in providing an explanation for irregular migration in one of the cities considered, given the important differences displayed by the phenomenon in the two contexts, the same argument will probably fail to explain the phenomenon in the other. If, for instance, the high number of irregular migrants in Madrid and their relatively easy daily experience were explained as the result of the inefficacy of the Spanish state, and, on this basis, it was claimed, more in general, that irregular migration evidences the failures of states to control populations, the low numbers of irregular migrants in Amsterdam and their very difficult conditions would then contradict such conclusion. This example shows how both points of view have problems and become contradictory if used to simultaneously explain two geographically different cases. Yet, similar problems arise also if one of the two arguments is applied to explain irregular migration in a single case, but when different chronological moments are considered. For instance, if the large number of Ecuadorian irregular migrants in Madrid and their relatively comfortable living conditions until 2008 are interpreted as a signal of state failure, the drastic reduction of numbers in the subsequent years and the deterioration of migrants' living conditions would suggest the opposite. In general, then, it seems possible to conclude that both theoretical perspectives fail to provide explanations for irregular migration that are able to withstand a comparative test.

Yet, even laying aside the comparative ambition, how effective are the explanations provided by the two theoretical perspectives if applied to the single cases? Let us first see the "performance" of the state failure thesis. Although the irregular migration realties experienced by Ecuadorian irregular migrants in Amsterdam and Madrid are very different and certainly evidence a different level of effectiveness on the part of the Dutch and the Spanish states to control irregular migration, it is frankly difficult to consider any of the two states as unable to control the phenomenon. A good indication of this is given by the fact that, in both cases, the political interventions regarding the migration regime put forward with the explicit intention of tackling the irregular migration phenomenon, in the Netherlands in 1998, in Spain in 2004, have unequivocally shown a degree of efficacy, determining changes that clearly emerged in the experience of Ecuadorian irregular migrants. In Amsterdam, after 1998, Ecuadorian irregular migrants faced a drastic reduction in the work opportunities and had increasing difficulties in accessing healthcare services and finding permanent accommodation. In Madrid, after 2004, Ecuadorian migrants found it increasingly difficult to enter Spain irregularly and, if already in the country, they were strongly discouraged from maintaining their irregular status. As it emerges, though many of the forces and processes signalled by researchers as weakening states abilities to control population have been found at work both in Amsterdam and Madrid, the two cases have also shown, at the same time, successful efforts on the part of states to regain control and to increase the efficacy of their policies.

Taking now into consideration the choice thesis, the idea that states intentionally create or allow irregular migration, either to satisfy internal political needs (or hidden agendas) or to please societal demands, also the explicative capacity of this thesis appears to be rather limited. If neither in Amsterdam nor in Madrid states seem to have lost control on the migratory dynamics, it is also true that in neither case do they seem to be able to completely control the phenomenon. In this respect, the Amsterdam case is particularly telling. Despite the great, prolonged, and seemingly systematic efforts made by the Dutch state to fight irregular migration, Ecuadorian migrants in Amsterdam, although with increasing difficulties and deteriorated conditions, have been able to adapt to the changing conditions and to find ways to realize, at least in part, their aspirations. This fact shows that the "countervailing power" represented, for instance, by migrant's agency or by the role of the civil society can all but be underestimated. Another aspect inherently related to the choice thesis that results quite problematic is the assumption that states are monolithic, almighty actors that are able to design and implement fully coordinated and coherent interventions. As it emerges from the data analysed in Chap. 6 and from migrants' accounts in Chap. 7, despite noteworthy differences, both the Dutch and the Spanish states display, on the one hand, important administrative, budgetary and logistic limitations in their ability to deal with irregular migration, and, on the other hand, high degrees of internal complexity (levels and sectors of government, administrative organization, policy construction and implementation procedures, etc.) which determine policy incoherencies or even contradictions. In relation to this, the experience of controls Ecuadorian migrants underwent in Amsterdam and Madrid, is particularly revealing. Notwithstanding the very different approach adopted by the two states – the Dutch more discreet but pervasive, the Spanish more "spectacular" but unsystematic – in both cases migrants were able to find inconsistencies, for instance, legal loopholes, a lack of coordination, implementation weaknesses, and, taking advantage of these, find strategies to circumvent controls.

As pointed out in the critical discussion of both failure theories and choice theories of irregular migration, at the end of Chap. 3, the explicative limitations of these theories, confirmed by the collected data, were related to three fundamental conceptual weaknesses: a problematic understanding of society, often assumed as subsumed to the state; a problematic understanding of social actors, both institutional and individual, imagined as monolithic, coherent, and time stable; and a problematic understanding of social interactions, interpreted in deterministic, cause/effect terms. The data emerged from the empirical study appears to reinforce this interpretation. Regarding the first, both in Amsterdam and Madrid, the experience of Ecuadorian irregular migrants appears to be the dynamic result of very complex interactions among social actors, each behaving according to their own interests and logics. Within this scenario, although with significant differences between the two cases, states have emerged as certainly crucial actors, probably the most influential, but in neither case are they able to fully control other social actors and determine the overall social outcomes. Accordingly, to make sense of the characteristics of the irregular migration phenomenon in the two cities what needs to be analysed is not states but a larger entity that exceeds them and includes all social actors and their interplay, i.e. society. In relation to the second conceptual weakness, the picture offered by the two analysed cases also evidenced the internal fragmentation and circumstantial variability of each social actor. Challenged by the complex societal dynamics, both institutions and individuals have been scarcely able to produce single, fully coherent, time consistent reactions. More often their responses have appeared as partial and never fully settled mediations among the multiplicity of interests, desires and logics present within each of them and an even more complex set of interests, desires and logics present in the social environment. This aspect has been particularly evident when considering institutional actors and in particular states. The "image" of these type of institutions that was possible to be drawn from the analysis of their policies and the experience Ecuadorian irregular migrants had of them in Amsterdam and Madrid, is that of extremely complex assemblages of relatively autonomous subcomponents, active at different levels, in different sectors, with different functions, each according to specific internal logics. Although to describe states' actions it is common to use terms such as "state will", "state decision" or "state intentions (real or covered)", these operations must be understood as a useful expedient to simplify communication. The uncritical acceptance of the analytical implications of this type of communication, however, is problematic because it determines a transfiguration of reality. Finally, focusing on social interactions, a similar problem was evidenced. In the "social arena" represented by Amsterdam and Madrid, no actor, not even the states, appeared to be able to perfectly asses all variables and on that basis design and implement actions capable of determining direct, fully predictable cause-effect relations. What it was possible to observe, instead, was a very blurred picture. Each actor, on the basis of their own, inevitably partial, understanding of reality proposes strategies intended to achieve desired results. The implemented actions, however, trigger complex environmental dynamics, the sum of the other actors' responses, that can lead to extremely variable degrees of success. As was possible to observe, the limited effects of many of the policies aimed at controlling irregular migration, were partially determined by the reactions triggered by those actors, irregular migrants in the first place, interested in minimizing their effects.

8.1.2 Ecuadorian Irregular Migration in Amsterdam and Madrid: An Attempt to Explain It Through a Social Systems Perspective

As just discussed, one of the main weaknesses of the "classic" theoretical approaches to irregular migration is their difficulty to explain the phenomenon as a differentiated one. Accordingly, using such approaches, it was impossible to come up with a convincing, comprehensive explanation of the empirical study's results, which, in Chap. 6, showed a very different structural context affecting migrations in the cities of Amsterdam and Madrid, and, in Chap. 7, a very different lived experience of Ecuadorian irregular migrants in the two cities. Now, is it possible to advance an explanation of irregular migration that is able to account for such a differential picture? In other words, is it possible to produce a theory of irregular migration that is able to successfully deal with its differentiation?

The discovery of important differences with regard to both contexts and the shape that irregular migration takes within them may appear rather obvious or somehow meaningless. It is not. As suggested by the systemic theoretical approach developed in Chap. 4, this result forces one to assume social complexity and difference as the starting point when approaching the irregular migration phenomenon. No context is equal to another; no irregular migrant lives under the same conditions as another. Assuming complexity and difference as the starting point, however, does not mean that comparisons and generalizations are not possible, that it is necessary to simply accept that everything is unique and therefore not comparable to the rest, that, since the whole context is different, the only possible explanation for the characteristics of a certain phenomenon is the difference of the whole context, ultimately, that a theory of irregular migration is impossible. What it does mean is that comparisons and generalizations, in order to effectively offer elements of analysis, can work at a higher level of abstraction. Therefore, for instance, it makes sense to compare the overall social condition of irregular migrants within a certain context or degree of restrictiveness of the migration regime, even more than the specific experience of a single migrant or a particular migration control action or law. What it also means is that linear, definitive monocausal explanations, such as those proposed by the "classic" theoretical approaches, must be abandoned in favour of systemic explanations that assuming complexity as the starting point, are able assess the different influence of each factor. In this sense, the practice of searching for explanations for irregular migration becomes a hermeneutic exercise, perhaps less definitive in its conclusions, but certainly closer to the complexity of reality. Bearing this in mind, in this section a systemic explanation of irregular migration in Amsterdam and Madrid will be proposed.

Recalling the conclusions of Chap. 4, two main ideas are important. Firstly, irregular migration should be understood as a complex, differentiated, structural phenomenon of modern world society. The development of this phenomenon should be related to the existing structural mismatch between the dominant form of social differentiation (functional) and the specific form of internal differentiation (segmentary) into territorial states of the political system. This creates a fundamental conflict between two logics: the all-inclusive logic of most social systems (economic, legal, educational, familial, etc.) that fosters human mobility across geographic space, and the exclusive logic of states that insist on regulating human mobility on the basis of a membership principle. Against this backdrop, irregular migration emerges as an adaptive solution to the mismatch existing between the demand for entry into certain states and the limited number of legal entry slots available. Secondly, if, in abstract and theoretical terms, irregular migration is explained as a structural feature of world society, the concrete, sociological manifestations embodied by the phenomenon within specific contexts must be empirically researched and subsequently explained as the result of a context-specific, dynamic, evolutionary interplay among: (A) functional social systems; (B) states; and (C) migrants. As suggested by the theory of social systems, moreover, each of these actors needs to be considered as autopoietic, self-referential and internally differentiated and social relations must be interpreted through an irritation/resonance model instead of an input/output model.

Then, if in abstract terms, the different irregular migration realities in Amsterdam and Madrid can be explained as the specific, context-determined, embodiment of world society's structural mismatch between the social demand for entry and the limited number of slots available, to explain the concrete, sociological manifestation of the phenomenon, it is necessary to explore and interpret the specific interrelation that in each city exists between the characteristics of the structural context and those of the related irregular migration reality. The first step in order to achieve this goal will be to "distil" the results of both the context analysis and the fieldwork in order to produce more abstract comparable results. Once one has these more abstract results in hand, the proper exercise of interpretation will be endeavoured.

Contexts

At the end of Chap. 6, a synoptic comparison of a number of important structural characteristics affecting the irregular migration phenomenon in the cities of Amsterdam and Madrid was presented. In particular, it was compared: the historical trends of migration, the migration regimes, the economies, the states structures and capacities, the public opinion and political stance regarding migration. In Table 8.1, it is possible to observe the results of the context analysis and an attempt at abstraction of them. By distilling and combining the role of the different elements, it seems possible to locate three main abstract comparable structural features that have affected, although in dissimilar ways, irregular migration both in Amsterdam and Madrid. Each of these general features can be understood as the result of a combination of the effects of a number of others and can vary along a continuum between two poles.

Table 8.1 Abstracting for structural contexts: Amsterdam (read from left to right) and Madrid (read from right to left)

Amsterdam →				Madrid ←			
Migration trends	Historical	• Old country of migration.	From moderate, to low	From very high, to moderate, to low	• Recent country of migration.	Historical	Migration trends
	Irregular migration	• Moderate until 2002, low afterwards.			• Very high until 2005, moderate until 2007, low afterwards.	Irregular migration	
Migration Regime	Legal channels	• Narrow labour migration channels. • Broad asylum seeker channels.	Highly restrictive	Inconsistently restrictive until 2004 Selectively open after 2004	• Narrow labour migration channels (until 2004); Flexible labour migration channels (from 2005). • Narrow asylum seeker channels.	Legal channels	Migration Regime
	Regularization	• Very sporadic and limited regularizations • No permanent regularization schemes.			• Recurrent, massive regularizations • Available permanent regularization schemes.	Regularization	
	Internal controls	• Strict after 1998 • Irregular migrants excluded from healthcare after1998 • No public spaces controls. • Improved deportation policy	Medium until 1998, strict after 1998 Enforcement: increasingly strict the 2000s	Weak until 2005, medium after 2005 Enforcement: increasingly strict after 2005.	• Increasingly strict after 2005 • Irregular migrants excluded from healthcare since 2012 • Public spaces controls in spells • Improved deportation policy	Internal controls	
Economy, labour market, shadow economy	GDP	• Booming economy 1994-2000 and 2006-2007 (GDP over 2.5%). • Mild economic crisis since 2009.	Low demand of unskilled workers	High demand of unskilled workers until 2008 Low demand of unskilled workers after 2008	• Booming economy1995-2007 (GDP over 2.5%) • Deep economic crisis since 2009.	GDP	Economy, labour market, shadow economy
	Labour market	• Slow growth of total employment. • Unemployment: stable, very low unemployment			• Creation of jobs between 1998 and 2007 (+6.5 millions). Destruction of jobs after 2008 (-3.4 millions) • Unemployment: decreasing until 2007; steeply rising afterwards.	Labour market	
	Sectors	• Limited low-skilled sectors			• Important low-skilled sectors.	Sectors	
	Shadow econ.	• 14% to 9%			• 24% to 19%.	Shadow econ.	
Welfare State	Type	• Conservative	Medium to high regulation capacity and social interventionism	Low regulation capacity and social interventionism	• Southern/Mediterranean	Welfare Type	Welfare State
	Main principles	• Social insurance + Social assistance			• Social insurance	Main principles	
	% of GDP	• Between 25% and 30%			• Between 20% and 25%	% of GDP	
	Main universal services	• Education, Healthcare, Old age pensions, Old age assistance			• Education, Healthcare, Old age pensions	Main universal services	
	Labour market controls	• Increasingly strict along the 2000s			• Increasing after 2005	Labour market controls	
Politics, public opinion, migration	Anti-immigration discourses and political parties	• Increasing importance of anti-immigration discourses in public and political debates. • Anti-immigration parties.	Anti-immigration discourses Anti-immigration political vectors	No anti- immigration discourses No anti-immigration political vectors	• No anti-immigrant discourses at a national level. • No anti-immigrant parties.	Anti-immigration discourses and political parties	Politics, public opinion, migration
	Public concern over migration	• High to moderate.			• Low to moderate.	Public concern over migration	
	Integration policies	• Increasingly restrictive policies since 1990s.			• Liberal policies until 2012; then increasingly restrictive.	Integration policies	

A. *The social demand for unskilled labour*. This feature combines the effects of the economic trends (GDP variation), the labour market structure and the size of the shadow economy. It can vary between a high demand, usually discernible in connection to growing economic trends, segmented labour markets and sizable shadow economies, and a low demand, often associated to stable or decreasing economic trends, unified and regulated labour markets and reduced shadow economies.
B. *The migration regime*. This feature combines the effects the public opinion and political attitude towards irregular migration, migration history, political culture, external influences (international institutions) and the existence of political vectors of anti-immigrant discourses. This feature can vary between a restrictive and non-restrictive configuration. The first pole is usually associated to long or complicated migration histories, strict political cultures, to possible restrictive imperatives from partners or supra-national institutions and to the presence within society of active and successful vectors of anti-immigrant discourses. The second pole can be connected to recent or relatively non-conflictive migration histories, flexible or more relaxed political cultures, no external restrictive pressures and the absence in society of anti-immigration vectors.
C. *The political system's capacity* to regulate and influence social transactions in relation to the other social systems (economy, law, religion, communication, etc.). This feature combined the effects of: history, political culture, and the extension, efficiency and culture of administration. This feature can vary between a preponderant and a subordinate political system. The first pole is usually associated to a longer and more successful history of the political system's organizations, to stricter, more legalistic and statistic political cultures, to older, more efficient and strict administration cultures. The second pole is more often discernible in cases of more recent and less developed political system's organizations, more fragmented and "private" political cultures and to younger, less efficient and more flexible administration cultures.

In Amsterdam, the social demand for unskilled labour appears to have been moderate in the 1990s and slightly diminishing from then on. This can be related to the relatively stable economic trends, the limited low-skilled sectors, the size of the shadow economy. Its continuous decrease in recent years can be linked to a further reduction of the underground economy and the effects of the economic crisis. Regarding the migration regime, a clearly restrictionist trend has been observable. The legal channels have been reduced and tightly controlled, the internal control policy has been reinforced and a policy of exclusion of irregular migrants has been enforced. This can be related to anxieties within the public opinion and the political attitude towards migration, which can also be connected to the migration history and the crisis of the Dutch integration model. Finally, concerning the capacity of the political system, it seems possible to consider this as medium to high. A number of elements support this claim: the level of intervention in the social transactions (sectors, level of expenditure, regulation of the labour market), the level and continuous reduction of the shadow economy, the increasing effectiveness in the expulsion policy.

In Madrid, the social demand for unskilled labour appears to have been very high until 2008 and very low afterwards. This can be related to the combination of: the economic trends, extremely positive in the first phase and the very opposite in the second one; the structure of the labour market and the importance of sectors such as construction, care work and private house cleaning; the weight of the shadow economy. Regarding the migration regime, although with some internal contradictions and a slightly restrictive trend, this has been characterized by the availability of regularization channels, the low levels of internal controls and the inclusion of irregular migrants. Finally, concerning the capacity of the political system, it seems possible to consider this as low to medium. This can be related to the lower level of intervention in the social transactions (sectors, level of expenditure, regulation of the labour market), the importance of the shadow economy, the lower efficiency in internal control policies.

Irregular Migration Realities

At the end of Chap. 7, a synoptic comparison of the results of the fieldwork was presented. The main features of the experience of Ecuadorian irregular migrants in Amsterdam and Madrid were compared, in particular: the legal trajectories and regularization channels, the experience regarding work, internal controls, housing and healthcare. In Table 8.2, it is possible to observe the results of the analysis of irregular migration realities and an attempt at abstraction. Distilling and combining the role of all the elements it appears possible to locate three main abstract features that can be compared and that have characterized the experience of irregular migrants, although in different ways both in Amsterdam and Madrid. Each of these characteristics combines the effect of others and can vary along a continuum between two poles.

A. *The size of the irregular migration population*. This feature can vary between a large or small irregular migration population.
B. *The average length of the irregular migration experience*. This feature is determined by: the availability of *ad-hoc* regularization channels, the availability of alternative channels and the elaboration of strategies to regularize. It can vary between long-term irregular migration and short-term irregular migration.
C. *The life conditions of irregular migrants*, determined by: the availability and conditions of working opportunities, the experience of internal controls and the fear of deportation, the accessibility to housing and to healthcare assistance. This feature can vary between good and bad living conditions.

In Amsterdam, the size of the irregular migration population appears to have been relatively significant at the end of 1990s, when it represented almost 30% of the total foreign population. Down through the 2000s, this proportion substantially

Table 8.2 Abstracting from irregular migration realities: Amsterdam (read from left to right) and Madrid (read from right to left)

Amsterdam →						
Irregular population size				*Before mid-2000s* • Medium (27%-1998)	*After mid-2000s* • Small (15%-2009)	**Medium to Small**
Length	**Legal trajectories**			• Never regular (17/39) • Regularized through marriage or cohabitation agreement (11/30) • Regularized under exceptional circumstances (2/30)		**Long term**
	Average irregular years			• 12 years		
	Regularization Options			• No ad-hoc channels, difficult.		
Conditions	**Work**	Availability		*Until 2005* • Medium	*After 2005* • Low	**Medium to Low**
		Sectors		• Many	• Very few	
	Controls	On work sites		*Until mid-2000s* • Medium	*After mid-2000s* • High	**Medium and systematic to Strict and systematic**
		Public spaces	Direct	• Not available	• Not available	
			Indirect	• Medium	• High	
		Deportation fear		• Medium	• High	
	Housing			• Very difficult		
	Healthcare			*Until 1998* • Easy access	*After 1998* • Very difficult access	**Difficult access**

← Madrid						
Very High to Small	*After 2007* • Small (20%-2009)	*Before 2007* • Very high (50%-2003)				**Irregular population size**
Short term	• Regularized using ad-hoc channels (21/30) • Befallen irregularity (6/30) • Never regular (3/30).				**Legal trajectories**	**Length**
	• 5 years				**Average irregular years**	
	• Ad-hoc channels, easy				**Regularization options**	
Very high to Very low	*After 2008* • Very low	*Until 2008* • Very high		Availability	**Work**	**Conditions**
	• Very few	• Many		Sectors		
Low and unsystematic to Variably strict and unsystematic	*After 2004* • Medium	*Until 2004* • Low		On work sites	**Controls**	
	• High in spells	• Low	Direct	Public Spaces		
	• Medium	• Low	Indirect			
	• Medium	• Low		Deportation fear		
	• Easy				**Housing**	
Easy access	*After 2012* • Easy access	*Until 2012* • Difficult access			**Healthcare**	

fell; the last available data for year 2009 showed that the irregular migration population represented less than 15%. It is important to recall that this reduction occurred without the adoption of massive regularization processes. Regarding the average length of the irregular migration experience, this can be considered as long-term. The data that emerged from the fieldwork showed that most migrants could not find ways to regularize their status and that they had been living irregularly for an average of 12 years. Concerning the living conditions of the irregular migration population, this can be considered as increasingly tough since the end of the 1990s. Although the working conditions have been generally good, the availability of opportunities has fallen. Controls have become stricter and stricter in the working sites and the fear of deportation has substantially increased. Housing remains one of the most complex problems for irregular migrants in Amsterdam, the accessible options being usually expensive and unstable. Finally, the access to healthcare has become severely restricted to irregular migrants since 1998.

In Madrid, the size of the irregular migration population was very substantial in the first years of the 2000s, when it represented more that 40%. This proportion sharply fell after 2005, and has maintained a decreasing trend since then. Regarding the average length of the irregular migration experience, this can be considered as short-term. The data that emerged from the fieldwork showed that most migrants were able to regularize their status thanks to the existence of many *ad-hoc* channels, and had lived irregularly for an average of 5 years. Concerning the living conditions of the irregular migration population, this can be considered as good until 2008 and increasingly hard afterwards. The working conditions and the availability of opportunities were very positive until 2008. From that year on, the circumstances abruptly changed: it became very difficult to find employment (especially for men), wages fell and options were usually unstable. Controls were very limited until the second half of the 2000s. After 2008 and especially during certain periods, there were raids in public spaces, like metro stations, buses, parks and bars. Except for the very first years, housing has not been a major problem for Ecuadorian irregular migrants in Madrid. Access to healthcare was free for irregular migrants until 2012.

Assessing Systemic Relations

Adopting a systemic perspective, the relation between contexts and irregular migration realities appear complex, dynamic and multi-causal. The specific characteristics that the experience of Ecuadorian irregular migrants had in Amsterdam and Madrid appear as difficult to be deduced from a single factor and explained in its relation. On the contrary, what emerges is the existence of eco-systems made of different components, which interact and influence each other, creating the condition for the irregular migration phenomenon to appear and evolve. In Table 8.3 it is possible to observe the parallel evolution of contexts and irregular migration realities in the two cities.

Table 8.3 Assessing systemic relation

Amsterdam				Madrid			
Context		Irregular migration reality		Irregular migration reality		Context	
Social demand for unskilled labour	Moderate To Low	Medium To Small	**Irregular population size**	**Irregular population size**	Big To Small	Very high To low	**Social demand for unskilled labour**
Migration regime	Restrictive To Highly restrictive	Long-term	**Status Length**	**Status Length**	Short term	Inconsistently restrictive To selectively open	**Migration regime**
Political system capacity	Medium To High	Medium To Hard	**Conditions**	**Conditions**	Easy To Hard	Low To Medium	**Political system capacity**

The most evident relation that has surfaced from the analysis of the Amsterdam case is the one between the implementation of an increasingly restrictive migration regime since the end of the 2000s and the toughening of the conditions for irregular migrants. As discussed, such change can be understood as part of anxieties within the public opinion and the political attitudes towards migration. The reduction of entry channels, exclusion of irregular migrants from healthcare and other social services, the stricter controls on labour and the implementation of a more efficient exclusion policy may have certainly contributed to the reduction of the irregular migration population. Yet, this result may have not been attained without a political system that was able to efficiently implement its policies and deeply penetrate different spheres of the social life (labour market, housing market, identification and expulsion policy). At the same time, the relatively low demand for unskilled labour, resulting from the sectorial structure of the labour market and the reduction of the shadow economy, have certainly been relevant as well. In this sense, it is possible to say that the restrictive effects that polices have evidently had on the lives of irregular migrants, have been possible within the context of a more general systemic structure that has favoured this outcome.

The most evident relation that emerged from the analysis of the Madrid case is the one between the sharp change in the social demand for unskilled labour after the start of the economic crisis at the end of 2007. Even if a number of reforms to the migration regime in 2004 and the adoption of massive regularization had certainly contributed to the reduction of the stock of irregular migrants, the effect of the sudden and deep change in the labour market had a deep impact on the conditions of irregular migrants. In the space of one year, the working opportunities became very limited, especially for men, salaries decreased and the jobs became extremely precarious. Interestingly, the modification in the labour market affected also the possibility of irregular migrants to regularize their status or to renew their residence permit in the case they had already got one. In this sense, the dynamics of the economic system reverberated through other important aspects of migrants' lives.

Moreover, it is possible that the adoption of very "spectacular" control measures, such as the raids in public spaces and the amendment to the free healthcare-access policy may have also been a result of the economic downturn. The reduction of the administration budget may have favoured the adoption of a less organic and less costly control policy and the attempt to reduce the budget by reducing rights. Also in the case of Madrid, then, it seems possible to recognize a sort of systemic reaction that, originating from one of the systems, the economic one, has determined a reaction that has involved the other systems.

Coming to a conclusion, in general terms, the different irregular migration realities existing in Amsterdam and Madrid can be explained in relation to the particular shape that the structural mismatch between the social demand of migrants – determined by the interplay of all social systems – and the limited number of entry slots available – determined by states – assumes in the two cities. Yet, the analysis of the relations existing in each case between the characteristics of the structural context and those of the irregular migration reality, allows one to take a step further. In particular, it allows one to put in relation the effects of specific "macro-structural" features that characterize the context – each of these the sum of several "micro-structural" features – to specific features of the experience of irregular migrants.[1] In Amsterdam, irregular migration has been a numerically reduced phenomenon, characterized by the long-term period of the status and the difficult social conditions for migrants. This characteristics can be linked to: a structurally limited demand for unskilled labour; the existence of a highly restrictive migration regime; and a pervading political system capable of largely (not fully) controlling social transactions and therefore effectively implementing its policies. In Madrid, irregular migration has been a highly variable numerical phenomenon, characterized by the short-term length of the status and variable social conditions. This characteristics can be linked to: a highly variable demand for unskilled labour; a rather open migration regime; and a non-pervading political system capable of controlling social transactions to a lesser extent and therefore moderately effective in the implementation of policies.

8.2 Further Steps Towards a Systemic Theory of Irregular Migration

8.2.1 An Analytical Framework for Irregular Migration

On the basis of results obtained and the discussion proposed in the previous section, in this final section, a typology of context/irregular migration reality relations will be presented. The objective is not to create a fixed structure of causal relations or a deterministic tool of analysis but to propose a tentative scheme of analysis, a hypothetical

[1] The proposed distinction between "macro-structural" and "micro-structural" features only refers to the fact that the former are the cumulative result of the sum of the latter.

space of interactions that allows one to visualize recursive relations between structures and irregular migration phenomena. Such an analytical tool is coherent with the social systems' perspective outlined in Chap. 4. Having assumed that no social actor (not even states) is able to fully determine social outcomes, that every actor is internally complex and that social interactions work through an irritation/resonance model, the point is not to produce, once again, linear explanations or to identify decisive actors (that either succeed or fail). The objective is to design an instrument, complex enough to be able to acknowledge the role played by a multiplicity of actors, to take into consideration the dynamics of actions and reactions that each of them trigger, to account for the different degrees of success and failure that such dynamics entangles but, at the same time, sufficiently simple to foster the identification of possible regularities, stronger linkages and recurrent patterns.

The area of the proposed scheme is divided into 8 spaces by three axes (see Fig. 8.1). Each axis represents one of the three main structural features affecting irregular migration that have been identified – the social demand for unskilled labour, the migration regime, the political system' capacity – and the variability of that feature between two poles. Each of these macro-structural features is the cumulative result of the effects of other micro-structural characteristics. In Table 8.4, there is a schematization of the main influences that was possible to observe (others may be possible).

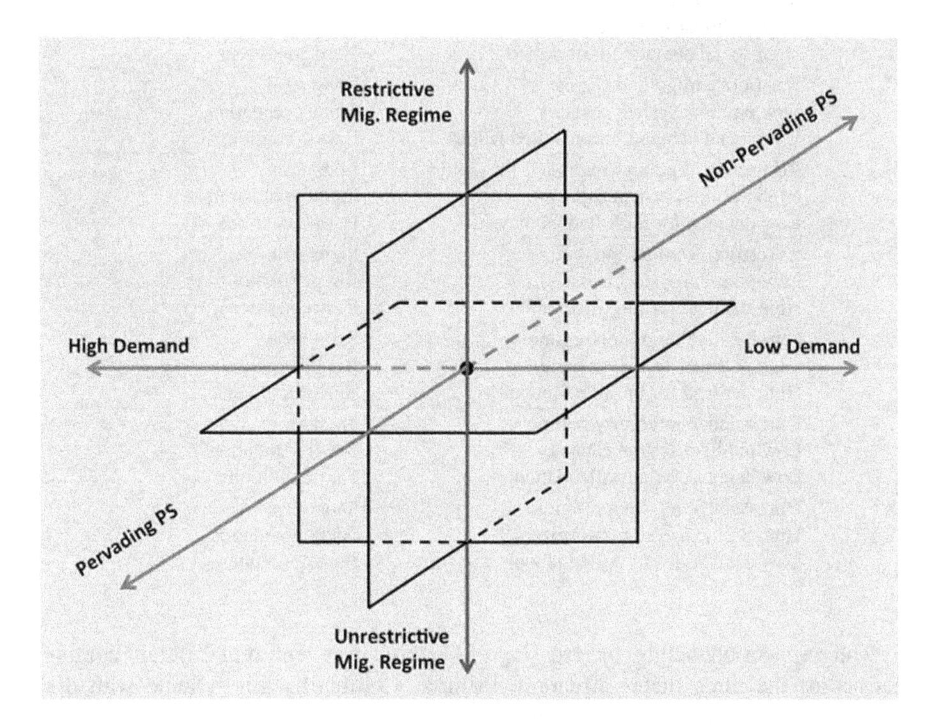

Fig. 8.1 Structures and irregular migration realities

Table 8.4 Context features that affect irregular migration

Macro-structural features	Micro-structural features
Social demand for unskilled labour	Economic cycle: expansion/contraction. Structure of the labour market: degree of segmentation and size of the informal sector Size and networks built by migrants Geopolitical position of sending and receiving countries Welfare state regime
Migration regime	Migratory background and cultural attitude towards migration Demographic cycle: percentage of migrant population, number of migrant generations, ethnic conflicts Politicization of migration: high/low
Political system capacity	Political culture: statist/non-statist. Implementation capacity: high/low. Regime type: liberal/non-liberal. Administrative culture and tradition

Table 8.5 Irregular migration typology

	Structural context	Irregular migration realities
1	Restrictive migration regime Low political system capacity High social demand for unskilled labour	Long term Big population Easy conditions
2	Restrictive migration regime High political system capacity High social demand for unskilled labour	Long term Big population Hard conditions
3	Restrictive migration regime Low political system capacity Low social demand for unskilled labour	Long term Small population Easy conditions
4	Restrictive migratory regime High political system capacity Low demand for unskilled labour	Long term Small population Hard conditions
5	Unrestrictive migration regime Low political system capacity High demand for unskilled labour	Short term Big population Easy conditions
6	Unrestrictive migration regime High political system capacity High demand for unskilled labour	Short term Big population Hard conditions
7	Unrestrictive migratory regime Low political system capacity Low demand for unskilled labour	Short term Small population Easy conditions
8	Unrestrictive migratory regime High political system capacity Low demand for unskilled labour	Short term Small population Hard conditions

The 8 spaces created by the crossing of the three axes represent different combinations of the three macro-structural features. Combining this scheme with the results of the Amsterdam/Madrid comparison, it is possible to imagine 8 types of irregular migration (see Table 8.5). Each of these types represents an ideal-type that links the combination of structural conditions to the characteristics of the irregular migration reality. They describe an overall general tendency, the type of irregular migration produced by a certain social configuration. It is clear that important exceptions are possible and that to fully describe the experience of each migrant, it is necessary to add an analysis of each individual trajectory. To give an example, space 4 represents a structural context with a restrictive migratory regime, a high political system capacity and a low social demand for unskilled labour. This is a context that prompts an irregular migration reality characterized by long term irregular statuses, small populations and hard conditions. Reconnecting to the results of the empirical study, this combination could represent well the situation found in Amsterdam in the most recent years considered in this study. Space 5, on the other hand, represents a structural context with an unrestrictive migration regime, a low political system capacity and a high social demand for unskilled labour that translates into an irregular migration reality marked by short term irregular statuses, big populations and easy conditions. This is a combination that could represent well the situation found in Madrid before the effects of the economic crisis in 2008.

The proposed analytical framework is coherent with the idea of irregular migration as a differentiated phenomenon that emerges from the particular configuration that the different social systems maintain within a specific context. According to this perspective, the irregular status of a migrant, by itself, does not tell much about the social, lived experience of this migrant. In each context, such status translates into a number of opportunities and limitations that can be extremely different. The analysis of irregular migration realities, at the same time, reveals important characteristics of the context in which they emerge, the dynamic equilibrium between different social systems, the existence of a certain systemic coherence that affects the evolution of all the different social systems.

8.2.2 Study Strengths and Limitations

The combination of different theoretical and empirical research strategies, and the attempt to establish a dialogue between them, has offered interesting material and original viewpoints from where to question the existing theoretical explanations for irregular migration and explore other possible perspectives. In particular, the systemic theoretical approach to irregular migration, based on Niklas Luhmann's social system theory, appears to have offered a stimulating and effective alternative capable of overcoming many of the limitations displayed by more "classic" approaches. Interestingly, such advantages, that had been prefigured in the theoretical discussion, were confirmed once the theory was tested in relation to the data which emerged from the empirical study. Not only was the systemic approach able to withstand the comparative challenge, offering an explanation of irregular migration capable of making sense of the realities that surfaced within very different contexts, such as Amsterdam and Madrid, but it was also able to offer a more realistic, multi-causal account of such realities.

The analytical framework of irregular migration presented at the end of this chapter, which represents the most advanced step towards a comprehensive systemic interpretation of possible irregular migration realities, is, at the same time and necessarily, the weakest. Although, the proposed idea – that of a schematization capable of assessing the cumulative impact of the different contextual characteristics in terms of some, decisive macro-structural features and of relating these to the characteristics of the irregular migration realties that in each context emerge – appears encouraging, because it offers a flexible, non-reductive tool of analysis. In order to confirm its usefulness, and, if that is the case, to produce more sophisticated, strengthened versions of it, it is crucial to expand and diversify the comparative research in this field. In particular, it would be necessary: to comparatively explore the experience of irregular migrants of different nationalities within the same contexts, in order to assess the impact of this variable; to compare the experience of irregular migrants in other countries that present similar structural conditions to those already considered, to observe possible differences among similar cases; to consider other countries with very different structural contexts, in order to further explore the insights of the "most different case" strategy.

Having considered its many limitations, the data presented in this book is significant and its discussion in relation to the proposed systemic theoretical approach appears to have contributed to the affirmation of an innovative perspective in the understanding of one of the most complex and characteristic phenomena of our time. The road to cover is long but the suggested route seems promising.

Printed in the USA
CPSIA information can be obtained
at www.ICGtesting.com
LVHW021934181223
766698LV00009B/1653